Sources for the Study of Science, Technology and Everyday Life 1870–1950
Volume 1: A Primary Reader

Sources for the Study of Science, Technology and Everyday Life 1870–1950

Volume 1: A Primary Reader

edited by
Gerrylynn K. Roberts
at the Open University

HODDER AND STOUGHTON
LONDON SYDNEY AUCKLAND TORONTO
in association with the Open University

British Library Cataloguing in Publication Data

Sources for the study of science, technology
and everyday life 1870–1950.
1. Science, ca 1850–1976 2. Technology,
to ca 1950
1. Roberts, Gerrylynn K. II. Chant, Colin
III. Open University
509′.034

ISBN 0–340–49002–0

First published in Great Britain 1988

Typeset in Times by Gecko Ltd., Bicester, Oxon
Printed and bound in Great Britain for Hodder and Stoughton Educational, a division of
Hodder and Stoughton Limited, Mill Road, Dunton Green, Sevenoaks, Kent by Richard
Clay Ltd, Bungay, Suffolk

Contents

Preface

The selection of primary source extracts and illustrations in this volume was made by the authors of an Open University second-level course, *Science, Technology and Everyday Life, 1870 – 1950*. It is one of three core course books which will be used throughout the students' year of work.[1] The diverse nature of the extracts from chapter to chapter reflects differences in individual authors' pedagogic styles in deploying the varied elements available, as well as differences in the topics themselves. However, each of the extracts and illustrations is directly relevant to the course, which explores a range of issues in the social history of technology in relation to the changes of the 'Second Industrial Revolution'.

Although this book has been assembled especially for the use of Open University students, other readers concerned with the social history of technology in our period should find much of interest. The volume can hardly pretend to be definitive of the subject, but is nonetheless rich in its offerings. The course concentrates on the main technological innovations connoted by the term, 'Second Industrial Revolution', as can be seen from the chapter headings. To cope with so immense a canvas, little is said about the persistent effects of the *First* Industrial Revolution, nor about innovations with roots in the period but social effects principally outside it. The range, however, from electric light to penicillin is still enormous.

Where possible, biographical information is provided about each of the extract authors, but no introductory analyses are offered, so as not to pre-empt their pedagogic use. Considerable editing of the extracts has been necessary, but all elisions are marked. It remains only to thank colleagues on the course team and secretarial colleagues as well for their assistance in preparing this volume.

Note

1 Chant, C. (ed.) (1988) *Science, Technology and Everyday Life 1870–1950*. London: Routledge and Kegan Paul; Chant, C. (ed.) (1988) *Sources for the Study of Science, Technology and Everyday Life: Volume Two, A Secondary Reader*. London: Hodder and Stoughton.

Acknowledgments

The editors and publishers would like to thank the following for permission to reproduce material in this volume:

Associated Book Publishers UK Ltd for the extract from *The Martyrdom of Man* by William Winwood Reade, published by Routledge & Kegan Paul; *British Medical Journal* for 'DDT in General Use', 1946 and 'Penicillin' by H. Florey, 1944; Constable Publishers for the extract from *Electric Cooking, Heating, Cleaning; A Manual of Electricity in the Service of the Home* by Maud Lancaster, 1914; Duckworth & Co. for 'The Microbe' from *More Beasts for Worse Children* by Hilaire Belloc, 1937; Electrical Review for 'Presidential Address to the Institution of Electrical Engineers' by S. Z. de Ferranti, published in *The Electrical Review*, 67, 1910; W. H. Freeman and Company for 'The Food Problem' by Lord John Boyd-Orr, published in *Scientific American*, 183, August 1959; David Higham Associates Ltd for the extract from *Man's Conquest of Nature* by F. Sherwood Taylor, 1948; The Institution of Electrical Engineers for 'The EAW All Electric House', 1936; Her Majesty's Stationery Office for 'Report for 1948–49' by the Ministry of Health, Great Britain, 1949, published by permission of HMSO; Johns Hopkins University Press for 'What Was the Niagara Central Station Plan?' by C. F. Scott, published in *Networks of Power: Electrification in Western Society, 1880–1930* by T. P. Hughes; Macdonald for the extract from *Workman's Cottage to Windsor Castle* by John Hodge; Macmillan Magazines Ltd for 'Presidential Address to the British Association for the Advancement of Science' by Sir Alfred Ewing, reprinted by permission from *Nature* Vol. 130, 1932, copyright © 1932 Macmillan Magazines Ltd; Macmillan Publishing Company for the extract from *English Sanitary Institutions Reviewed in their Course of Development, and in some of their Political and Social Relations* by John Simon, 1890; Oxford University Press for the extract from *The Century of Hope: A Sketch of Western Progress from 1815 to the Great War* by F. S. Marvin, reprinted by permission of Oxford University Press; Oxford University Press for 'Frozen Food' by Siegfried Giedion, published in *Mechanization Takes Command: A Contribution to Anonymous History*, 1969; Progress Publishers for 'The Immediate Tasks of the Soviet Government' by V. I. Lenin, published in *Collected Works*, 1965; The Society of Chemical Industry for 'Problems Arising from the Disposal of Effluents Containing Synthetic Detergents', published in *Chemistry and Industry*, October 1949 and 'Plastics' by H. V. Potter, published in *Chemistry and Industry*, 8 March 1941 and 'Nylon Yarn' by G. Loasby, published in *Chemistry and Industry*, 5 August 1944; The United States Government Printing Office for 'Transportation' by H. A. Osgood, published in *Technological Trends and National Policy including the Social Implications of New Inventions* and 'Communication by Wire and Wireless' by T. A. M. Craven and Committee, published in *Technological Trends and National Policy including the Social Implications of New Inventions*; Watkins/Loomis Agency Inc. for the extract from *Men and Machines* by Stuart Chase, published by Macmillan New York, 1929.

The publishers would also like to thank the following for their permission to reproduce illustrations:

Plate 1: Punch; *Plate 2* (left): shot from Mikhail Amikst (ed.) *Soviet Commercial Design of the Twenties*, trans. Catherine Cooke, reproduced by permission of Thames & Hudson/John Calmann & King Ltd; *Plate 2* (right) New York World's Fair Corporation; *Plate 3*: The Electricity Council; *Plate 4*: The Institution of Electrical Engineers; *Plate 5*: BBC Hulton Picture Library; *Plate 6*: London Transport; *Plate 8* (above): Geffrye Museum; *Plate 11* (left): Museum of English Rural Life, Reading University; *Plate 11* (right): H. J. Heinz & Co. Ltd; *Plate 12* (above): Fox Photos; *Plate 12* (below): Hunting Aerofilms; *Plate 13* (above): University College London; *Plate 13* (below): The Mansell Collection Limited; *Plate 16* (left): The Wellcome Institute Library; *Plate 16* (right): BBC Hulton Picture Library.

Every effort has been made to trace and acknowledge ownership of copyright. The publishers will be glad to make suitable arrangements with holders of the following items whom it has not been possible to contact:

Plate 7 (left and right); *Plate 9*: from *A Million and One Nights* by T. Ramsaye, published by Frank Cass & Co. Ltd/Simon and Schuster, 1926; *Plate 10*; *Plate 14*: from *Household Engineering: Scientific Management in the Home* by Christine Frederick, published by George Routledge, 1919; *Plate 15* (above and below) from *From the American System to Mass Production 1800–1932* by D. A. Hounshell, published by Johns Hopkins University Press, 1984.

1

Science, Technology and Progress

1.1 William Winwood Reade, *The Martyrdom of Man*, 1872

[. . .]

The upper classes in America have not that exquisite refinement which exists in the highest circles of society in Europe. But if we take the whole people through and through, we find them the most civilised nation on the earth. They preserve in a degree hitherto without example the dignity of human nature unimpaired. Their nobleness of character results from prosperity; and their prosperity is due to the nature of their land. Those who are unable to earn a living in the east, have only to move towards the west. This then is the reason that the English race in America is more happy, more enlightened, and more thriving, than it is in the mother-land. Politically speaking, the emigrant gains nothing; he is as free in England as he is in America; but he leaves a land where labour is depreciated, and goes to a land where labour is in demand. That England may become as prosperous as America, it must be placed under American conditions; that is to say food must be cheap, labour must be dear, emigration must be easy. It is not by universal suffrage, it is not by any act of parliament that these conditions can be created. It is Science alone which can Americanize England; it is Science alone which can ameliorate the condition of the human race [. . .]

We can conquer nature only by obeying her laws, and in order to obey her laws we must first learn what they are. When we have ascertained, by means of Science, the method of nature's operations, we shall be able to take her place and to perform them for ourselves. When we understand the laws which regulate the complex phenomena of life, we shall be able to predict the future as we are already able to predict comets and eclipses and the planetary movements.

Three inventions which perhaps may be long delayed, but which possibly are near at hand, will give to this overcrowded island the prosperous conditions of the United States. The first is the discovery of a motive force which will take the place of steam, with its cumbrous fuel of oil or coal; secondly, the invention of aerial locomotion which will transport labour at a trifling cost of money and of time to any part of the planet, and which, by annihilating distance, will speedily extinguish national distinctions; and thirdly, the manufacture of flesh and flour from the elements by a chemical

Source: William Winwood Reade, *The Martyrdom of Man*, London: Kegan Paul & Co., 1972, pp. 510–515.

process in the laboratory, similar to that which is now performed within the bodies of the animals and plants. Food will then be manufactured in unlimited quantities at a trifling expense; and our enlightened posterity will look back upon us who eat oxen and sheep just as we look back upon cannibals. Hunger and starvation will then be unknown, and the best part of the human life will no longer be wasted in the tedious process of cultivating the fields. Population will mightily increase, and the earth will be a garden. Governments will be conducted with the quietude and regularity of club committees. The interest which is now felt in politics will be transferred to science; the latest news from the laboratory of the chemist, or the observatory of the astronomer, or the experimenting room of the biologist will be eagerly discussed. Poetry and the fine arts will take that place in the heart which religion now holds. Luxuries will be cheapened and made common to all; none will be rich, and none poor. Not only will Man subdue the forces of evil that are without; he will also subdue those that are within. He will repress the base instincts and propensities which he has inherited from the animals below; he will obey the laws that are written on his heart; he will worship the divinity within him. As our conscience forbids us to commit actions which the conscience of the savage allows, so the moral sense of our successors will stigmatize as crimes those offences against the intellect which are sanctioned by ourselves. Idleness and stupidity will be regarded with abhorrence. Women will become the companions of men, and the tutors of their children. The whole world will be united by the same sentiment which united the primeval clan, and which made its members think, feel, and act as one. Men will look upon this star as their fatherland; its progress will be their ambition; the gratitude of others their reward. These bodies which now we wear, belong to the lower animals; our minds have already outgrown them; already we look upon them with contempt. A time will come when Science will transform them by means which we cannot conjecture, and which, even if explained to us, we could not now understand, just as the savage cannot understand electricity, magnetism, steam. Disease will be extirpated; the causes of decay will be removed; immortality will be invented. And then, the earth being small, mankind will migrate into space, and will cross the airless Saharas which separate planet from planet, and sun from sun. The earth will become a Holy Land which will be visited by pilgrims from all the quarters of the universe. Finally, men will master the forces of nature; they will become themselves architects of systems, manufacturers of worlds. Man then will be perfect; he will then be a creator; he will therefore be what the vulgar worship as a god. [. . .]

1.2   F. S. Marvin, *The Century of Hope: A Sketch of Western Progress from 1815 to the Great War*, 1919

[. . .] We are not [. . .] discussing any short cut to peace or any substitute for a League of Nations, but inquiring, purely in an historical spirit, what tendencies may be discovered in recent years towards an ideal which all admit to be desirable, though many doubt its near approach or even its

Source: F. S. Marvin, *The Century of Hope: A Sketch of Western Progress from 1815 to the Great War*, 2nd ed., Oxford: Clarendon Press, 1919, pp. 337–40.

possible attainment. [. . .] Human skill and perseverance in piercing the St Gothard, human insight and synthesis in tracing the curves and learning the constituents of the most distant stars, human care and ingenuity in analysing disease and chasing the poisonous bacillus from the blood, the noble human emotion, in all its compass and gamut, which speaks in a symphony of Beethoven—these things are the true uniting forces; and, as a rule, in recording the achievements of the past, we put these in the smallest type or leave them out altogether. But they have been growing all the while, and the nineteenth century was their best flowering-time. [. . .] It is, of course, true—so obvious in these days that it is scarcely worth mention—that the railway may be a strategic weapon to bring troops through Belgium, that the airplane may be employed in dealing death to thousands, and the wireless convey instructions to contending fleets. These have been the portents of the war, and their warlike value happens to quicken invention and stimulate use. But, broadly speaking and looking to the future as well as to the past, the effect of the industrial inventions and scientific applications of the last century has been something quite different from war, not wholly good but certainly unifying. The analogy of a great state, thus somewhat superficially unified, may throw some light upon the larger problem. Russia had been linked up by a transcontinental railway, by telegraphs and telephones, as well as by a common system of law and administration. But the national soul was not fully awake. The disasters of war broke into this, and for the moment we see a chaos of conflicting chiefs, hostile parties, and petty nationalities. But does any one suppose that the previous unification, the material and mechanical links, will go for nothing? Is it not certain, on the contrary, that the actual unity achieved, imperfect though it was, was a fact of permanent importance and that the ideal of a 'great Russia' will remain to modify and bring together in some new form the congeries of smaller units which are arising from the wreck?

It remains, however, profoundly true—the most important fact in our whole discussion—that the spiritual forces, of which we may trace the workings in the same period, are the supreme factors, both in building the individual soul and in giving a common soul to all humanity. This common spirit is best exhibited, and most powerfully enlarged, in the two channels of the growth of science and the application of science, especially in the art of medicine. We put these first not from any theory of their intrinsic worth. The inspiration of poetry or music is another, it may be a higher, thing than the unfolding of the mysteries of matter or the growth of the living cell. But in the history of science and its applications we have the most perfect example of a growing human product in which the diverse races of mankind have all taken a proportionate share as they advanced in civilization. There has been absolute similarity in the mental process and a growing solidity in the accomplished structure. In this region national differences are simply irrelevant, and personal jealousies merely a mark of personal inferiority. Now in this sphere, the international development of science and of medicine, using both terms in the widest sense, the nineteenth century, remarkable for so much, was most remarkable of all, for its advances were more than those of all the earlier centuries together. It is no fanciful analogy, but the closest approximation we can make to the truth, to say that this scientific structure, established and being taught at the close of the century in all Western countries, corresponded in its general relation to life, in the respect which it inspires in its students, in the number and international union of its teachers, with the mass of mediaeval theology and philosophy which was to Dante the sum of human knowledge. [. . .] It is in this sphere, the sphere of pure intellect, as Dante showed, the unity of mankind is most fully realized. All seats of learning, whether universities or learned societies, or associations for spreading knowledge in wider circles, are in reality the organs of a true internationalism, and strengthen the human spirit by knowledge springing from a universal source and tending ultimately to the universal good.

1.3   Stuart Chase, *Men and Machines*, 1929

Slaves and Philosophers

Certain philosophers hold that machinery is enslaving us. I am not a machine tender, but first and last I encounter a good many mechanisms in a day's march, particularly when that day is spent in a city so large and so complicated that it could never have been built by human muscle. Before analyzing the extent of serfdom in others, it might be well to determine how far I am myself a slave.

The first thing that I hear in the morning is a machine—a patented alarm clock. It calls and I obey. But if I do not feel like obeying, I touch its back, and it relapses humbly into silence. Thus we bully each other, with the clock normally leading by a wide margin. (Once, however, I threw a clock out of the window, and it never bullied anyone again.)

I arise and go into the bathroom. Here I take up a second mechanism, and after inserting a piece of leather between its rollers, move it briskly up and down before proceeding to scrape my face with it. I turn various faucets and a mixing valve, and a nickel dial studded with little holes showers me with water. Depending on the season, I may snap on electric lights and an electric heater. Downstairs, if it chances to be either the first or the fifteenth day of the month, I take a can with a very long nose, and oil an electric motor which blows petroleum into my furnace, a motor which runs the washing machine, and a motor which operates my refrigeration engine. Meanwhile an electric range is cooking my breakfast, and on the table slices of bread are being heated by an electrical toaster which makes a buzzing sound in its vitals, and then suddenly splits open when the toast is browned to a turn. If time allows, I may play a little tune on the piano which stands near the breakfast table, noting the delicate system of levers and hammers upon which the mechanism is based. Before I leave the house, the whine of the vacuum cleaner is already in my ears.

I go to the garage, and by proper and sometimes prolonged manipulations, start explosions in six cylinders of an internal combustion engine. With foot and hand, I put the revolving crank shaft in touch with the rear wheels and proceed to pilot the whole mechanism to the station, passing or halting before three sets of automatic signal lights as I go. At the station, I cease operating machinery and resign myself to another man's operation of an enormous secondary mover, fed by a third rail

Source: Stuart Chase, *Men and Machines*, New York: Macmillan, 1929, pp. 1–6.

from a hydroelectric turbine at Niagara Falls. I cannot glance out of the window without seeing a steamboat on the Hudson River, a steam shovel on the speculative real estate development, a travelling crane on a coal dock, or a file of motor cars on any street. Every so often comes the faint roar and silver glint of an airplane, winging its way above the river.

Arrived at the metropolitan terminal, I buy a package of cigarettes by depositing a coin in a machine which hands me matches and says, 'Thank you; it's toasted.' I then spend ten minutes walking just three blocks. If I tried to shorten this time appreciably, I should most certainly be killed by a machine. Instead, I look down into an enormous pit where the day before yesterday, according to the best of my recollection, there stood a solid brownstone house. Now it is an inferno of swarming men, horses, trucks, pile drivers, rock drills, steam shovels, clacking pumps, and preparation for erecting a gigantic steel derrick. From across the street comes the deafening rat-tat-too of riveters.

I enter my office building and a machine shoots me vertically towards the roof. I step into a large room, stopping for a moment on the threshold to sort out the various mechanical noises which lend a never-ending orchestral accompaniment to all my working hours in town. The sputter of typewriters; the thud as the carriage is snapped back; the alternate rings and buzzes of the telephone switchboard; the rhythmic thump of the adding machine; the soft grind of a pencil sharpener; the remorseless clack of the addressograph and the mimeograph. During the day I make and receive about twenty calls upon the telephone. I crank an adding machine from time to time. I may operate a typewriter for an hour or so. Meanwhile my eye can seldom stray long from my watch, if the day is to be got through with at all.

To go up or downtown I use one of the three horizontal levels of transportation which the city affords. As a profound melancholia always accompanies a trip on the lowest, I endeavour to use the upper two exclusively. Many of my fellow citizens do the same, particularly since a score of them were killed at Times Square the other day. Killed in the rush hour, like beeves in the Chicago stockyards; except that the packers put no more animals into a pen than can go in.

In the evening I reverse the morning process. At home, I may sit for a few moments beneath a machine which gives off ultra-violet rays, or I may dance to strains of a machine which runs a steel needle over a corrugated rubber disc, and for the governor of whose delicate mechanism we are indebted to James Watt. For days at home, direct contact is limited to running the motor car and making minor repairs upon it; answering the telephone; using, hearing, tinkering with the various household so-called labour-savers—particularly the plumbing system.

In the summer, by way of contrast, I may spend weeks in a mountain camp, where the only mechanisms are the motor car, the telephone, and a remarkably temperamental contrivance for pumping water. Year in year out I doubt if my direct contact with machines averages much over two hours a day. When I go to town, the ratio runs considerably higher; when I stay at home, an hour would certainly cover it; in the summer, an hour would be too much.

So far as I am aware, no permanently evil effects befall me by virtue of these two mechanical hours. I suffer from no prolonged monotonies, fatigues or repressions. The worst moments are dodging street traffic and hearing its roar, riding in the subway, changing tires and cleaning out the incinerator. When the telephone becomes unduly obstreperous, I go away and leave it. By far the most fatiguing noise in my office is the scraping of chair legs on the hard composition floor—and chairs I believe antedated Watt. All the depressions that I suffer from direct contact with machinery are certainly compensated for by the helping hand it holds out to me—a calculator for figuring percentages, an oil heater which requires no stoking, a reading lamp which does not have to be trimmed and filled, an elongated radius of travel possibilities, a car for errands, together with the genuine thrill which often comes from controlling its forty horses.

I do not feel like a slave, though of course I may be one all the same. Clocks and watches are hard masters but so they always have been; there is nothing new or ominous about their tyranny. No individual living in a social group is ever free, but I wonder if these two mechanized hours have put more shackles on me than were to be found on the average citizen of Rome two thousand years ago, or of China today—cultures innocent of engines both. As I look about the United States, the most mechanized nation under the sun, I have reason to believe—and later will bring in the statistical proof—that the number of those bound intimately to the rhythm of the machine is a small percentage of the total population, while there are probably more people with contacts remoter than mine than with closer contacts. In other words, I am more mechanized than the majority of my fellow citizens, and, needless to say, far less mechanized than a minority thereof. [. . .]

1.4   Sir Alfred Ewing, *Presidential Address to the British Association for the Advancement of Science*, 1932

In the present-day thinkers' attitude towards what is called mechanical progress we are conscious of a changed spirit. Admiration is tempered by criticism; complacency has given way to doubt; doubt is passing into alarm. There is a sense of perplexity and frustration, as in one who has gone a long way and finds he has taken the wrong turning. To go back is impossible: how shall he proceed? Where will he find himself if he follows this path or that? An old exponent of applied mechanics may be forgiven if he expresses something of the disillusion with which, now standing aside, he watches the sweeping pageant of discovery and invention in which he used to take unbounded delight. It is impossible not to ask, Whither does this tremendous procession tend? What, after all, is its goal? What its probable influence upon the future of the human race?

The pageant itself is a modern affair. A century ago it had barely taken form and had acquired none of the momentum which rather awes us to-day. The Industrial Revolution, as everybody knows, was of British origin; for a time our island remained the factory of the world. But soon, as was inevitable, the change of habit spread, and now every

Source: Sir Alfred Ewing, 'Presidential Address to the British Association for the Advancement of Science', *Nature*, vol. 130, 1932, p.349.

country, even China, is become more or less mechanized. The cornucopia of the engineer has been shaken over all the earth, scattering everywhere an endowment of previously unpossessed and unimagined capacities and powers. Beyond question many of these gifts are benefits to man, making life fuller, wider, healthier, richer in comforts and interests and in such happiness as material things can promote. But we are acutely aware that the engineer's gifts have been and may be grievously abused. In some there is potential tragedy as well as present burden. Man was ethically unprepared for so great a bounty. In the slow evolution of morals he is still unfit for the tremendous responsibility it entails. The command of Nature has been put into his hands before he knows how to command himself.

I need not dwell on consequent dangers which now press themselves insistently on our attention. We are learning that in the affairs of nations, as of individuals, there must, for the sake of amity, be some sacrifice of freedom. Accepted predilections as to national sovereignty have to be abandoned if the world is to keep the peace and allow civilization to survive. Geologists tell us that in the story of evolution they can trace the records of extinct species which perished through the very amplitude and efficiency of their personal apparatus for attack and defence. This carries a lesson for consideration at Geneva. But there is another aspect of the mechanization of life which is perhaps less familiar, on which I venture, in conclusion, a very few words.

More and more does mechanical production take the place of human effort, not only in manufactures but also in all our tasks, even the primitive task of tilling the ground. So man finds this, that while he is enriched with a multitude of possessions and possibilities beyond his dreams, he is in great measure deprived of one inestimable blessing, the necessity of toil. We invent the machinery of mass-production, and for the sake of cheapening the unit we develop output on a gigantic scale. Almost automatically the machine delivers a stream of articles in the creation of which the workman has had little part. He has lost the joy of craftsmanship, the old satisfaction in something accomplished through the conscientious exercise of care and skill. In many cases unemployment is thrust upon him, an unemployment that is more saddening than any drudgery. And the world finds itself glutted with competitive commodities, produced in a quantity too great to be absorbed, though every nation strives to secure at least a home market by erecting tariff walls. [. . .]

We must admit that there is a sinister side even to the peaceful activities of those who, in good faith and with the best intentions, make it their business to adapt the resources of Nature to the use and convenience of man.

Where shall we look for a remedy? I cannot tell. Some may envisage a distant Utopia in which there will be perfect adjustment of labour and the fruits of labour, a fair spreading of employment and of wages and of all the commodities that machines produce. Even so, the question will remain. How is man to spend the leisure he has won by handing over nearly all his burden to an untiring mechanical slave? Dare he hope for such spiritual betterment as will qualify him to use it well? God grant that he may strive for that and attain it. It is only by seeking he will find. I cannot think that man is destined to atrophy and cease through cultivating what, after all, is one of his most God-like faculties, the creative ingenuity of the engineer.

1.5   Hilaire Belloc, *The Microbe*, 1937

The Microbe

The Microbe is so very small
You cannot make him out at all,
But many sanguine people hope
To see him through a microscope.
His jointed tongue that lies beneath
A hundred curious rows of teeth;
His seven tufted tails with lots
Of lovely pink and purple spots,

Source: Hilaire Belloc, *More Beasts for Worse Children*, London: Duckworth, 1937, pp. 47–48.

On each of which a pattern stands,
Composed of forty separate bands;
His eyebrows of a tender green;
All these have never yet been seen—
But Scientists, who ought to know,
Assure us that they must be so. . . .
Oh! let us never, never doubt
What nobody is sure about!

1.6   F. Sherwood Taylor, *Man's Conquest of Nature*, 1948

The conquest of Nature we have so far made has had the practical results of fulfilling our physiological needs for food and health, decreasing pain, lengthening life and increasing our enjoyment by giving us more goods and services and more opportunity to enjoy certain of the arts. It could give us more leisure, but in fact has not done so. It has greatly increased our knowledge of physical things, and it has likewise enormously increased our power to do anything that can be accomplished by physical means.

How far, then, could this conquest of Nature, wisely applied, succeed in fulfilling the desires of man? Suppose that its success was such that everybody had enough to eat, a good education, a comfortable house, and a life of ninety years of perfect health during which only four hours a day were occupied in work; would not this be a good thing? Undoubtedly, a very good thing! But it would not be the completion of man's desires or a guarantee of happiness. Man can have all these and still quarrel with his wife, hate his children, be jealous of his neighbours, suffer from fits of boredom and depression, and be oppressed by the emptiness of human existence. He can still scheme to get more than he already has, however ample; amuse himself by seducing his neighbour's wife; band together with his fellows to subjugate some other human beings. In fact the conquest of Nature relieves man of none of his innate potentialities for experiencing and causing unhappiness. Indeed, anyone who has a circle of friends including both poor and wealthy is not likely to maintain that those who have a superfluity of goods and time are more happy and better members of the community than those who have not. Man's own nature is, in fact, at least as great an obstacle to his happiness as is external Nature.

Furthermore, man's nature being as it is, the conquest of Nature adds opportunities for its worst manifestations. Man desires to slay: here is the aeroplane and plutonium. Men wish to deceive other men into becoming their unconscious slaves: here is printing, transport, radio. Man wishes to save his fellows from disease and starvation: here is medicine, scientific agriculture, transport. It works both ways; but unfortunately it is much easier to break than to make, much easier to get short-term results by lies than by truth—so much so that today we are beginning to wonder whether the conquest of Nature is such a good thing after all. Obviously some advance from the primitive is to be desired; but obviously the world at present owes its devastation, at least as a cause *sine qua non*, to the conquest of Nature. It may be thought, then, that man is happiest when his degree of conquest of Nature is duly proportioned to his degree of conquest of his own mental nature. That conquest, in itself and alone, is capable of abolishing the misery that man would continue to suffer even if Nature were fully conquered.

[. . .] Yet we may wonder whether, in the event of man finding inner happiness, the conquest of Nature would be further pursued. For man, when at peace, finds in himself a sympathy with Nature unconquered and

Source: F. Sherwood Taylor, *Man's Conquest of Nature*, London: Paul Elek, 1948, pp. 111–114.

a power of understanding her, seeing upon her the sign manual of God's creation. Man's progress will be proportioned to his desires, and if the overmastering desire of man were the greater knowledge of God, I do not think he would worry very much about telephones and tin-openers. Civilisation would probably be greatly simplified, though it could never return to the disease-ridden squalor of the past. Whether science would progress it is hard to say, for man has hardly begun to consider the relationship between the knowledge that science gives and the knowledge of God. It may be that our study would be upon the knowledge of things not so much in their quantitative relationships as in their psychological significance—that natural science would relate to our whole knowledge of Nature and not merely to the shapes, sizes and movements of things.

2
Electrification

2.1   S. Z. de Ferranti, *Presidential Address to the Institution of Electrical Engineers*, 1910

[. . .]
It appears that with a problem such as we are discussing it is fundamental that the energy in the coal should be converted at as few centres as possible into a form in which it is most generally applicable to all purposes without exception, and in which it is most easily applied to all our wants, and is at the same time, in a form in which it is most difficult to waste it or use it improperly.

We are, therefore, forced to the conclusion that the only complete and final solution of the question is to be obtained by the conversion of the whole of the coal which we use for heat and power into electricity, and the recovery of its by-products at a comparatively small number of great electricity producing stations. All our wants in the way of light, power, heat and chemical action would then be met by a supply of electricity distributed all over the country. [. . .]

In the conversion of coal into electricity one of the most important considerations is the load factor at which the converting and distributing plant effects the operation.

Electricity used for lighting, cooking, power, and traction, must be supplied as and when required. On the other hand, domestic heating will be done largely through the medium of heat storage and is therefore a controllable form of demand. Metallurgical and chemical processes which depend for their success upon a very cheap supply of current, will have to be so adapted and modified that they can take current intermittently and so fill up the load curve, thus enabling the current which they require to be produced with the least capital expenditure, and, at the same time, greatly assisting the good conversion efficiency of the whole supply. I believe that under the circumstances a load factor of 60 per cent would be obtained. At first sight it seems unreasonable to expect such a load factor, but it must be remembered that our ideas are based on the present electric supply which only uses some 1 per cent of the coal now consumed in the country.

At present, as is quite natural, electricity is used for what coal does least satisfactorily direct, and it is misleading to compare the load factor so obtained with those which will be got when electricity replaces coal entirely. [. . .]

Source: S. Z. de Ferranti, 'Presidential Address to the Institution of Electrical Engineers', *The Electrical Review*, vol. 67, 1910, pp. 841–844.

The positions of the actual generating stations would be largely controlled by the facilities for obtaining coal and water for condensing. In many cases they would be close to the colliery districts, and the current would be transmitted to the points of demand. In other cases, where a considerable demand was concentrated at a distance from the sources of coal supply, and where the coal could be cheaply carried—especially by water—stations would be installed and would supply electricity to meet the surrounding demand. In all cases, however, whether far from, or near to the coal production, the coal would be delivered in very large quantities to only a few points of consumption, and would thus reduce the labour and cost of transmission and handling to the lowest figure.

Many works taking a large quantity of electricity for metallurgical and chemical purposes at a very low price would be built adjoining the generating stations, and other existing works would be at such short distances that the capital costs for distribution of the electricity which they used would be very small. [. . .]

It is interesting for a moment to consider the effect of such a supply of electricity upon its present and future uses.

Taking lighting to begin with, which was the first application for which a supply of electricity was generally given, it will be clear, considering the strong position which electric lighting now holds even with current at an average price of 2d. per Board of Trade unit, that when it is obtained from current at the much lower prices that would rule under the all-electric scheme, no other form of light would have a chance in competing with it.

Notwithstanding present high prices, a good deal of electric cooking and heating is already being used, and although it would appear to be too expensive for general application, still, the very good results obtained, and the large amount of labour saved, are already sufficient to justify its use to-day.

When electric heating and cooking are carried on with current at the very low figures at which it would be possible to sell for these purposes, it would only be a matter of time for all heating and cooking to be done by means of electricity.

Regarding the supply of power, electricity is now admittedly the most convenient form of power for all purposes, and this, again notwithstanding the costs involved on the comparatively small scale on which we now produce. The overwhelming advantages of electric power at a price at which it would be supplied on the all-electric scheme would clearly ensure its use for all power purposes.

The case with regard to electric tramways and light railways is well known, and any reduction in the cost of running due to cheaper current would, of course, act greatly in favour of these undertakings, and would help to extend their usefulness. Light railways, which, for various causes, have made such poor progress, if sensibly dealt with, would greatly benefit by finding a cheap supply of energy available in all parts of the country.

The electrification of main-line railways has not yet progressed very far, as it is hard to make out a sufficiently strong case to warrant the large expenditure necessary for electrification; but there is little doubt that growing traffic, which necessitates additional works, will be best met by electrification, which will enable a greater return to be obtained from existing lines and works. The electrification of our railways would be great-ly assisted and made a more profitable investment if a supply of current at such a figure as we are now considering were available for their working.

The manufacture of pig-iron is, no doubt, quite the most economical use of coal that we now have, but recent work with electric smelting furnaces has shown that it is only necessary to have electric current at a low enough price, and for sufficient experience to be obtained, to make it more eco-nomical to smelt iron electrically than by present methods, and using only sufficient coke to provide the carbon for the purpose of reduction. [. . .]

Steel-making electrically is already in extensive use, and even with present facilities for generating the current which the process requires, is beginning to make considerable headway. All steel would, of course, be produced electrically as soon as sufficient experience had been gained regarding details and a supply of very cheap current was available. Foundry work in both iron and steel would be most conveniently carried out by means of electric melting. It is already known that the electric furnace gives the best results obtainable for steel castings. The heating of steel for rolling, forging, and annealing will be most efficiently carried out electrically as soon as the cheap supply warrants experimenting in this direction. In fact, all furnace work for which coal or gas are now used could, I am convinced, be more satisfactorily done electrically when an abundant and cheap supply is available.

We now use aluminium for a number of purposes, notwithstanding our want of knowledge as to the best ways of working it. When our experience with aluminium in any way approaches what we now know about the working of steel, it is certain that vast quantities of this material will be used throughout the world. The manufacture of aluminium is another of the processes which will be greatly facilitated by a cheap supply of electricity. In fact, it may be said that aluminium can only be produced economically at present in water-power countries; but as an intermittent supply of electricity could be given under the proposed scheme at a lower price than it is being obtained at from water powers, we should be in a better position than the water-power countries; to manufacture this metal. With cheap electricity available, electro-chemical processes must grow and multiply to an enormous extent, and not only should we produce for ourselves all the chemicals which are now produced electrically abroad, but everything that can be produced electro-chemically would then be made in this country.

There is a further application of the electric current which, so soon as the price was low enough, would, no doubt, largely come into use. This is in the intensive growing of fruit and vegetables under glass. It is known that considerably more forcing in the way of heat can be advantageously applied where light is also furnished artificially, and it is therefore probable that, with electricity everywhere available at a low price, an immense amount of intensive cultivation under glass with the heat supplied by means of the electric arc would be undertaken, as in supplying heat by this means light would also be supplied, which would have the effect of enabling the growth to benefit fully by the artificial heat.

Summarising the whole position, it may safely be said that, whatever coal, gas, or power are now used, everything for which they are used will be better done when electricity is the medium of application.

Hardly less in importance in the all-electric scheme is the question of the by-products which become available by the proper use of our coal. These consist principally of fixed nitrogen, together with tar and oils. [. . .]

At present it is considered quite right and reasonable to canalise rivers and make great works for adding to the fertility of countries by means of irrigation, but I believe that in the future the time will come when it will be thought no more wonderful largely to control our weather than it is now thought wonderful to control the water after it has fallen on the land. I think that it will be possible to acquire knowledge which will enable us largely to control by electrical means the sunshine which reaches us, and, in a climate which usually has ample moisture in the atmosphere, to produce rainfall when and where we require it.

It seems to me that it may be possible, when we know a great deal more about electricity than we do to-day, to set up an electrical defence along our coasts by which we could cause the moisture in the clouds to fall in the form of rain, and so prevent these clouds drifting over the country between ourselves and the sun which they now blot out. It also seems to me that it will be possible, when more water on the country is required, to cause the falling of rain from the clouds passing over the highest part of the country and so produce an abundance of water which, properly used, would greatly add to the fertility of the country.

Of course, it may seem that these are only mad visions of the future, but I think we can hardly consider these results more improbable than anyone would have considered wireless telegraphy or flight in heavier than air machines 50 years ago. My excuse for mentioning these matters here is that they might constitute another great use of electricity, and their useful consummation would certainly be facilitated by an abundant supply of electrical energy.

There would be further by-products from the coal in the form of tar and light oils. The effect of their abundant production and sale at a low price would be most important to the country, as the large quantity of tar produced would enable us to make good roads, which we much need, and which would have the lowest cost of upkeep, and the light oils would, when carburetters have been further developed, go a long way towards supplying the fuel for our motor-cars and other motor-vehicles which we now have to import from abroad. [. . .]

Cheap electricity would greatly stimulate all manufacturing operations, which would, in turn, enable labour to be much better remunerated than at present, and to enjoy a much higher standard of comfort. The higher value of labour would in its turn stimulate inventiveness and the production of all sorts of labour-saving appliances which, with cheap electricity, would enable us to produce in the future under suitable market conditions at cheaper rates than are now possible, notwithstanding the better return that labour would obtain.

Great hardships are always produced where any great industrial change is made, but the more efficiently we can carry on the work of the country, the more margin must there be for the great majority of the people: so that any change which decreases the amount of labour required must eventually give the people greater comfort and less arduous work. It is hardly necessary to point out how much better the position of the country would be if we were producing the whole, or nearly the whole, of our requirements, as, in this case, we should be far less liable to be adversely affected by any external causes or by the occurrence of any great war.

At present, although the using of our coal may mean commercial activity, it certainly means the desolation of the country in parts where it is largely used. Instead of this harm being done to the country by our coal, we should fertilise the lands by its means and might even, as I have indicated, use it in the future to increase our sunshine.

Of course, there are many things which at present stand in the way of realising such a scheme as I have outlined. There are many technical details, which nothing but an immense amount of work can solve satisfactorily. There are also political and legislative difficulties standing in the way, but these, when the time arrived, would have to be got rid of rather than allow them to handicap the advance of the country. The more, however, that I have considered these ideas in detail, the more certain am I of the fundamental soundness underlying them and that it is only a matter of time before such a scheme is carried out in its entirety.

What interests us most, perhaps, is the question of how long it is likely to be before the all-electric idea becomes possible. At present there is so much required to be done to make it workable in all its details, that it seems as though its realisation would be long deferred. It must, however, be remembered that knowledge is continually being acquired which brings us nearer to its realisation, and that things engineering, and especially in electrical engineering, now move very rapidly. It may therefore come to pass that the all-electric idea, with its far-reaching changes and great benefits, will become an accomplished fact in the near future.

2.2 'Housewife' [Maud Lancaster], *Electric Cooking, Heating, Cleaning: A Manual of Electricity in the Service of the Home*, 1914

Introduction

The following pages are feeble efforts of mine to help my 'sisters in distress,' and to convince them of the wonderful blessings provided for us by nature's gift of Electricity which, aided by scientific research and inventions, is capable of doing so much towards bettering the home life.

Electric cooking and heating are by no means new, early attempts having been made so long ago as 1890, but partly on account of imperfect apparatus, but more by reason of the high prices which, until recently, were asked for electrical energy used for such purposes, their development and adoption have been slow.

Now, however, that apparatus has been perfected and specially favourable rates are available in most districts, there is nothing to hinder the widespread adoption of electricity, not only for cooking and heating,

Source: 'Housewife' [Maud Lancaster], *Electric Cooking, Heating, Cleaning: A Manual of Electricity in the Service of the Home*, ed. E.W. Lancaster, London: Constable, 1914, pp. 1–4, 24–27.

but for many purposes in the home at present carried out more or less successfully by hand. Once the simplicity, efficiency and perfection of electrical operation are realised, I am absolutely convinced that it will be adopted in every 'real' home.

Having been housekeeping for many, many years (too many to announce), I have unfortunately bought my experience and paid for it dearly, but 'as all things come to those that wait' I now revel in the bliss of an Electric Kitchen and electricity throughout the house generally. My home life, therefore, is much more easy, agreeable and healthful, both for my family, and my maids, and as servants or helps are like ourselves, 'human beings,' and good servants or helps are few, it is of the utmost importance to do all we can to make things healthful and easy for them, if we wish to ensure a placid and serene existence in our homes.

The advantage of Electric Lighting is now fully established and beyond question, but when its possibilities are more fully understood and applied, Electricity will be used for purposes unthought or undreamt of at present. Its use in *cooking, heating, ventilating, air purification* and *cleaning* marks the commencement of the electrical age, and I am convinced that it will soon become established in general use for these and many other purposes, and be looked upon as one of the greatest blessings in daily life in providing the home with *economic labour- and dirt saving service*—making existence for every wife and every maid or help more comfortable, more enjoyable and more healthful.

The hard-working husband also will find that things have changed for the better—for instead of finding on his return home, a 'neurotic' wife, worn out with the worries of housekeeping and domestic troubles, he will be welcomed by a loving woman, bubbling over with mirth and joy, a sure antidote for all the worries and trials which each man, more or less, has daily to encounter in this strenuous and competitive age.

Then, too, our little ones will be the happier, for we are apt to be so irritable, even with them, if our domestic arrangements are all upside down!

I do so wish to impress upon my readers *the vast importance* of our food *being properly cooked*. Good health is such a big 'factor' in the happiness of life, and it is wrong of us to neglect it! Digestion is impaired and ruined by overlooking this important fact, and I am convinced that if we, as 'housewives' (for, in spite of 'advanced ideas,' true housewives *do still exist*) will only devote a little time to the preparation of food for the sake of our own health and that of those around us, we shall save many of those dear ones from being semi-invalids, and aid in securing a race of more healthy and robust people.

In order to convince my readers that 'Electric Cooking' is the 'ideal method' and is likely to revolutionize all other systems, I have searched through various records of Ancient and Modern Cookery, and I cannot better substantiate my views than to quote from some of the reliable and scientific sources that have given me the data upon which I have worked with such great success! Generally there is no doubt that proper Cooking is *slow* cooking, and carried out electrically is absolutely *less costly* than by any other means, *apart* from the great saving of labour, the absence of dirt, and the better sanitary and hygienic conditions which accompany electrical operation!

Some of my readers may say, What have these historical facts, and the Chemistry of Cookery, to do with *Electric Cooking*! To these I say 'Everything'. These facts prove that it has been known for thousands of years that moderate, uniform and constant heat are the chief requirements in cooking. Those whom I shall name later on have been vainly trying to teach these principles, and others have been vainly trying to obtain from apparatus heated with coal, gas and other combustible materials, *a constant cooking temperature.* Owing to many causes, such as attention to fire, varying draughts, constant watching and turning of the article to be cooked, it has been impossible in an ordinary household to obtain the conditions necessary for this proper and hygienic form of cooking.

In this volume I deal chiefly with the application of such electrically operated appliances for use in the ordinary household as come chiefly under the sphere of woman's work.

Throughout all my interviews and demonstrations, which have been most instructive and helpful, I have kept absolutely to facts and statistics. May this little work, which is but a sketch of a most interesting and important subject, accomplish the end I have in view! Then I shall have the joy of knowing my efforts have not been in vain.

[. . .] The demand for electrical energy for power purposes is enormous, and in many districts exceeds that for lighting. This load has been built up by offering cheap units and it has proved a profitable source of revenue for supply undertakings. But, large as it is, it will be as nothing compared with the load furnished by domestic electrical requirements provided that energy be offered at reasonable and competitive prices so as to bring the advantages of electrical operation within the reach of all classes of society. Manufacturers will be only too glad to produce apparatus in large quantities at prices infinitely lower than is possible at present, and this will still further encourage the use of electrically-heated appliances.

The introduction of the metal-filament lamp, owing to its lessened current consumption, caused in many districts a serious drop in revenue from the lighting units sold. In one of the London districts for example, a drop of £7,000 was recorded from this cause. In other areas the loss of revenue has been proportionately serious, and in several districts the loss has been made up by a greatly increased consumption and by the addition of new consumers, attracted by the cheaper cost of lighting. In view of the smaller return from individual lighting installations, it has become necessary to encourage the use of devices other than those used for lighting, and the development of the domestic cooking and heating load opens up immense possibilities in this direction. If a householder already be a lighting consumer, no new service is required, and if some comprehensive system of charging be adopted, the need for separate circuits and extra meters in many cases can be avoided, so that the initial cost of providing for this class of business is quite small. It frequently does not pay to run an expensive service merely to supply a few tungsten lamps of low candle power, but if the consumer avails himself of a cooker or heater, the revenue from his installation, even though it be from cheaper units, is a paying proposition.

In America comparatively high rates have been charged for lighting supply, owing to the correspondingly heavy prices of gas, but many engineers are now catering for a heating and cooking load by reducing their tariffs to as low as 1½ cents per kelvin or unit, with a minimum monthly charge in some cases fixed at $1 only. In England, as stated elsewhere, 1d. or 2 cents per unit, is a fair average, and in upwards of 17 districts the rate is as low as ½d. or 1 cent per unit.

It is, however, not only in the matter of tariffs that the electric supply engineer must meet the consumer—there is the initial cost of the apparatus, and in many cases of the wiring to supply it. This difficulty can and must be surmounted by hiring and hire-purchase schemes. Any additional wiring can be dealt with on a rental basis, as is done with meters, so that the consumer will not be faced with a heavy bill at the outset for the cooking mains. Conditions naturally vary in each locality and they must be met by the supply engineer and tackled on their merits. The outlook justifies the facing of each problem in a spirit of optimism, and the advent of the all-electric age can be hastened, if every engineer sets out to cultivate the domestic load.

It should be the policy of every *Electric Supply Company or Committee* to SEE that its *Engineer has his own house equipped*, so as to get reliable data regarding the various apparatus. *The Central Station Engineer* (*and his Company or Committee*) must first believe in the possibilities of electricity for cooking and heating before he can hope to interest his consumers in the subject. Without personal knowledge and experience, as well as enthusiasm, little good can be accomplished. There are many supply engineers who consider electric cooking and heating as beneath their consideration, and as not offering a paying proposition, simply because they have not troubled to test for themselves in their own homes the advantages of electrical operation for uses other than lighting. Let *every station engineer* and sales manager set up *his own cooking installation and supervise the tests himself*. If he carries out his experiments in a businesslike and scientific manner, he cannot fail to gain implicit confidence, and so to be in a position to impress his consumers with a belief in electrical operation for the kitchen, for enthusiasm is infectious, and he will further be able to give first hand the results to his employers. No engineer should be content to wait for the demand for cooking and heating apparatus to come to him, but should take energetic steps to hasten and develop local interest in the subject, and this can most effectually be accomplished if he first becomes familiar, by personal experience, with the various domestic uses to which electricity may profitably be adapted. The field is enormous and if engineers will only take the trouble to go into the matter intelligently, and to study carefully *electrical cooking requirements* and the details of the various appliances on the market, they will realise that the golden opportunity has arrived and that the time is now ripe for pushing the use of electricity for many domestic uses undreamed of when lighting was the sole object of any electric supply undertaking. They must be alive to grasp and profit by the opportunity, realising that the wider applications of electricity will benefit not only their individual undertakings, but will alter for the good the lives of the people just as greatly as did the introduction of electric light and electric motive power.

The electrical press, both in Great Britain and America, is unanimous on the question. In America the *Electrical World* and in England the *Electrical Times* have especially identified themselves with the subject, and have done good work in bringing before the Central Station Engineer the true merits of the position. They have rightly pointed out the duty imposed upon him to see to it that his consumers avail themselves of all the advantages that electrical service can supply, and that it is of primary importance that he should familiarise himself first of all with every economical application to which electrical energy can be put in domestic and hotel life and industrial use. It is only a beginning to instal lamps for lighting in a house, although an important step in the right direction, and what is wanted is an educational campaign to bring home to consumers who merely use their installation for lighting that the good fairy of electricity can do greater things than these for them. The technical press is doing its level best to arouse engineers to the importance of the subject; the manufacturers are doing all and more than can be expected from them, and it remains for the supply engineer to use the material available and carry out the good work in his own district. Having taken off his coat to the work, he will have no cause to look back, and he can rely upon the cordial co-operation of everyone interested. There is ample scope for energy and enterprise; and a rich harvest, in the shape of improved load factors and increased outputs, awaits those engineers who have the courage to test these problems practically, *not* experimentally, and open up a campaign of combined publicity, on a commercial basis in favour of electricity for lighting, cooking, heating and the many other uses in every home.

2.3   Glasgow Corporation Electricity Department, *The Electric House: Some Practical Results*, 1925

The electric houses to be dealt with to-night are situated at Riddrie. Opened for public inspection in the spring of last year, the intense interest shown is evidenced by the fact that the daily average number of visitors was 1,500, or a matter of 28,000 over three weeks. Strange, is it not, that it is by an appeal to the eye that desire can best be created. Show the average woman over a house fitted with the latest electrical gadgets in actual operation, let her handle the electric iron, the vacuum sweeper, the cooker, and nothing will serve but she will eventually acquire them. The houses, equipped throughout with a complete range of electrical apparatus, were designed as an object-lesson on how to fit a modern labour-saving house. The absence of chimneys and fire-places was intended to demonstrate to the Corporation Housing Committee the savings possible by the installation of electrical heating and cooking apparatus. We endeavoured to prove that a modern electric house costs less to build and, what is perhaps more important, does not cost more to run than the average house with coal or gas fires.

The exhibit resulted, I believe in an awakening in many quarters to the possibility and desirability of electric service in the home. It focussed attention, created discussion, and undoubtedly helped forward

Source: Paper read before the Electrical Society of Glasgow by R. Hardie, reprinted as Glasgow Corporation Electricity Department, *The Electric House: Some Practical Results*, The British Electrical Development Association Pamphlet No. 553, February 1925.

our cause, the ultimate electrification of the City of Glasgow. As you may know it possibly assisted towards the decision of the Glasgow Corporation to erect three hundred houses, with electric light, electric cooker and wash boiler, and electric fires throughout. [. . .]

The late convener of the Electricity Propaganda Committee was sanguine enough to believe that once these electric houses are set going, gas need not be introduced for any purpose into any future building scheme. I share his confidence, and it is our intention to tackle the job in such a way that the electric houses to be built will remove the doubt which exists in the minds of many, even in the electrical trade, as to the possibility of running satisfactorily and economically an *artisan's* house by the aid of electricity.

It was thought desirable that the houses should be rented by members of the electricity department staff, by reason of the fact that they were for the first year or so to be open for inspection by interested visitors, and only those with a direct interest in the department were likely to put up with this inconvenience. The experiment was designed to prove to a sceptical public that it was possible to adopt electricity in a larger measure than was considered feasible for heating, cooking, and other purposes, at a cost well within the means of the bulk of our citizens, and to furnish *definite performance data* on the cost of operating such houses, so that we might all argue with a little more conviction for electric cooking and heating.

Equipment. The houses consist of living room, drawing room, three bedrooms, bath room, and kitchenette. It may perhaps be desirable to refresh your memories on the electrical installation, and I would emphasise the fact that I have only included apparatus in regular use.

First. The strictly 'all-electric house,' in which no coal is used. In the kitchenette there is the electric cooker, on which all the cooking is done. Electric hot water circulator which furnishes hot water throughout the house. A clothes wash boiler used on wash days once a week, and an electrically-operated clothes washing machine. An electric three-pint kettle in daily use, and an electric iron complete the equipment here.

In the living room or dining room a two k.w. electric fire, electric teapot and coffee percolator, toaster, food warming plate, cigar lighter. The dining room table is wired with three outlets, supplied from a plug in the centre of the floor.

Drawing Room. A two k.w. electric fire, standard lamps, piano lamp.

Bedrooms. Electric fires, hair dryer, vibrator, bed light fittings, night lights, and, perhaps most essential of all, the bed warmer.

Bath Room. Electric fire, towel rail, shaving cup.

Finally, the most cherished possession of the house, the electric vacuum cleaner.

The second house is identical, with the exception of one coal fire to heat the living room, and furnish hot water throughout the house. [. . .]

My experience is that the combination in use in my own house, i.e., a coal fire in the living room, to be used in the winter months, is what appeals at present to the average woman, and, after all, the woman's point of view, in domestic matters at least, is what counts. Tradition dies hard, and, while admitting the undoubted labour-saving and dirt-saving advantages of an electric fire in the drawing room, dining room, bedroom, or bath room, she still clings sentimentally to the blazing hearth in one apartment of the house. Ninety-nine per cent of the women you meet will argue that a coal fire is pleasanter to sit with. One advantage it has, of course, is that it burns the kitchen refuse, which is otherwise difficult to dispose of. This perversity or cussedness, as you might call it, on the part of the ladies, is to be deplored, because it retards the efforts being made to abolish the smoke nuisance in our large cities. It has long been recognised that the chief offender is the domestic chimney, and so long as the kitchen fire is retained we cannot have in its entirety the pure atmosphere we desire, even admitting that the general use of electric fires in other rooms of the house does ease the position somewhat. [. . .]

I am content to be judged by the facts. You may know that the Gas Department equipped with gas apparatus two houses at Riddrie, identical in size to ours, and the results published recently are of interest. [. . .]

Obviously, statistics can be made to prove anything, but my point is that our house was used without consideration of the fact that a test was being taken. But take the figures as they stand—10.7d. per day 'gas house,' 1s. per day 'electric.' Do you think the most cheeseparing, cantankerous, money-grabber would not deem the undoubted advantages and convenience of the electric house worth the difference of 1.3d. per day? There is no room for doubt on this score. For, let it not be forgotten that, in addition to the superior lighting, we had the use of vacuum cleaner, sewing machine motor, clothes washing machine—conveniences which are perforce denied to the occupants of gas houses. [. . .]

What hinders electrical development? Not heavy installation costs surely in these days of 20s. per point, and certainly if the figures quoted to-night mean anything, not the running cost. Mr Mitchell, in his opening address a year ago, emphasised the necessity of the broadening out of the demand in domestic premises. The reason is obvious enough. The consumer whose service lies out of action for most of the day all the year round and all day in the summer (as is the case with a lighting consumer) is not a paying proposition; and consequently, he directly retards the proper development of the electricity supply business.

This was brought home to the Glasgow Corporation in some of the recent housing schemes. Take, for instance, Mosspark, where, owing to the large capital expenditure and the meagre revenue to be expected, we stood to lose between £10,000 and £12,000 if electricity were used for lighting only. Other corporations have been faced with the same difficulty. A few of these schemes would result in the supply authority showing a loss on the year's working, or else would automatically mean keeping the over-all cost of current higher than it ought to be. Dundee in one housing scheme actually refused to accept lighting business only, and as a result the houses were lit by gas. Again, what hinders adequate development? Surely not any remissness on the part of the local supply authority. [. . .]

It may be desirable to give a summary of the various figures quoted by me:

1.	*Riddrie* 'All Electric'	£16	6	11
2.	*Riddrie* Part Electric	18	8	7
3.	*Riddrie* 'All Gas'			
	26 at 10s. 6d.; 26 at 5s. 3d., estimated	20	9	6
4.	*Riddrie* Part Gas, say, do.	14	0	0
5.	*Kirkhill* 'All Gas'	29	8	3

6. *Maryhill* Electricity for Lighting, Gas
    for Cooking, Coal for Heating   £18   1   10½

As I see it, the main lesson to be drawn from these figures is that electricity has nothing to fear from the closest investigation with regard to running cost.

It is admitted that *electric lighting* is without rival. The convenience and utility of *electric fires* were never questioned, and in to-day's improved models any objections as to unreliability and expensive maintenance have been swept away. The main obstacle to their widespread installation is the fact that they have still to be purchased outright, in addition to the cost of special wiring. A hiring scheme would at once get over this difficulty in existing houses; while in new property the cost of installing electric fires and the necessary wiring would be more than offset by the saving effected in doing away with open fireplaces and chimneys. We have still some way to go to convince architects that in a modern house the chimney is an expensive relic of the past, which in future building construction should be omitted, but we are undoubtedly making progress in this direction. [. . .]

2.4   Electrical Association for Women, *Electrical Outlet Campaign*, 1927

THE ELECTRICAL 'OUTLET' CAMPAIGN. Those who were responsible for bringing the Electrical Association for Women into existence had two definite objects in mind; the first, to provide facilities for the woman in the home to obtain a more intimate knowledge of the uses of electricity in our modern civilisation and especially in its relation to the home, and, secondly, to provide a platform for the expression of the woman's point of view on any question relating to electricity which may affect her private or public interests.

Up to the present time we have been mainly concerned with educational work amongst our members. This inaugural work has enabled our women to gain considerable knowledge of electrical economics and methods, and the time now seems ripe, with the knowledge thus gained, to begin to express the woman's point of view to the responsible authorities.

One of the most crying needs of the day is the lack of electrical 'Outlets' in the average house. In the existing house it is too often true that the unfortunate householder has to bear the cost of installation and to put up with the inconveniences which a little foresight in building would minimise or definitely anticipate.

The '*Electrical Outlet Campaign*,' which has been devised by the E.A.W., has for its object the provision of an adequate number of 'outlets' during building.

THE NEED FOR STANDARDISATION OF ELECTRICAL FITTINGS has also been realised by the E.A.W., and the new Campaign will urge upon the responsible electrical authorities the need for the use of standardised plugs and sockets for all domestic purposes. Everyone has experienced the annoyance of a badly fitting plug and socket and

Source: Electrical Association for Women, 'Electrical Outlet Campaign', *The Electrical Age*, vol. 1, 1926–1930, pp. 288–290, 383.

the difficulty of interchangeability.

Very good work in connection with the standardisation of plugs and sockets has already been carried out by the British Engineering Standards Association, which has issued a number of specifications for the manufacture of standardised fittings.

While our new Campaign is still only in its infancy, very definite and practical interest has already been shown in our efforts with promise of co-operation. The first practical result has been an invitation from the British Engineering Standards Association to appoint a woman representative to serve on the Electrical Accessories Committee of this Association to represent the woman's point of view.

How many 'Outlets' have you in your home?

When will the British housewife grow more impatient with the badly equipped houses in which she is expected to live and work and when will she cease to accept inefficiency as a necessary part of housekeeping?

We are told on all hands that electricity is to be the servant of the woman in the home, but how can we have a real electric labour-saving home unless we have the means whereby to use all those many useful appliances which are the real savers of labour.

Every modern woman has had the experience of going into an up-to-date Store and seeing displayed useful electric table fittings, vacuum cleaners, polishers and charming standard lamps; she may decide that these are within her purse limit, but then, having decided to purchase, she suddenly wonders how she is going to use them.

There is, perhaps, one lighting point and one heating point in the room, but then she may require to use the newly purchased coffee percolater at the same time as the electric light and electric radiator are also in commission.

True, there are numerous adaptors on the market which can be used, but how much better if in each room there were provided a number of 'Outlets' or, as they are sometimes called, 'Wall Plugs'?

How much easier it would be to plug in the vacuum cleaner, the polisher and the extra standard lamp exactly where they are needed.

With the coming of the warm weather the prudent housewife may decide to invest in a refrigerator, but on further reflection she will ask herself the question 'Where can this be plugged in?'

Now the Government has decided that the country is to be provided with an adequate supply of electricity, and so has brought in the Electricity (Supply) Act, 1926, which has already commenced to operate in certain parts of the country.

Great provisions have been made for the erection of huge mains in the form of a grid, super-power stations are to be erected and, in fact, all the important questions relating to electrical generation and transmission have been carefully worked out. A supply of electricity, however, will be of little avail to the women of the country if the houses are not already provided with the facilities for using this new power.

This point we consider to be the *weak link* in the otherwise very successful chain forged by the Government to help the housewife to more efficient home-making.

A campaign to strengthen this weak link has, therefore, been devised by the E.A.W. to be known as 'The Electrical Outlet Campaign,' which will demand that in all new housing estates, provision shall be made for the future labour-saving home by providing an adequate electric service during building.

The campaign will include visits to housing estates throughout the country to find out what provision is being made for an adequate electric service, and later deputations will be arranged to architects and housing authorities in different parts of the country.

In order that some indication may be given to the housing authorities of the woman's point of view in this matter, the E.A.W. is drawing up a 'National Woman's Specification' showing the number and position of electrical 'Outlets' which it is necessary to provide for each room in the house.

Already our campaign has been described in the Press and elsewhere as a *practical commonsense idea*, and we feel that this effort on our part will do much to ensure that the housewives of the future will have a better equipped and more efficiently conducted house than those we find to-day. [. . .]

Women's national specification

As mentioned in the Editorial Notes the first step of our Campaign has been reached in the publication of the Women's National Specification.

Our readers will remember that the Electrical Outlet Campaign is being undertaken by the Electrical Association for Women: 'To ask that during the building of New Houses provision shall be made for the future labour-saving home by the installation of an adequate electric service.'

The Electricity (Supply) Act of 1926 was passed by Parliament to hasten the day when the great servant of Electricity could be brought within the reach of every home-maker in the country.

This Supply will be of little avail if the house is not already provided with the facilities for using this new power. In other words there must be sufficient 'Outlets' already installed in the home.

It is cheaper and neater to wire a house for an electrical service during Building than after the house has been erected.

A preliminary survey of Houses now in course of erection shows that very little provision is being made for 'Outlets.'

Hence the need and justification for the Women's National Specification.

WOMEN'S NATIONAL SPECIFICATION

1 This Specification has been drawn up as a result of discussions at Headquarters in London and amongst the Branches at Glasgow, Birmingham, Manchester, Cheltenham, Newcastle and Cardiff.

2 The discussions have been based on the plan of a house as approved by the Ministry of Health, and represents the number of 'Outlets' which the Association considers necessary for an Electrical Labour-Saving Home.

The Association does not assume responsibility for the *Type* of house, but it was necessary to have a plan of an ordinary standard house on which to base discussion hence the use of this plan.

(Attention is drawn to the fact that if the house had been designed for an All-Electric Service, considerable reduction in cost of building could have been effected by the elimination of chimneys, etc.).

3 The Association urges that 'Outlets' in new houses should be of standard design and dimensions as fixed by the British Engineering Standards Association, provision being made for earthing in the bathroom, kitchen and scullery. Necessary provision for safety should also be made throughout the house whether by earthing or any other method.

It is desirable that 'Outlets' in the kitchen should be waist-high and sunk in the wall.

4 The size and current capacity of 'Outlets' must depend upon individual and local circumstances, but stress must be laid on the necessity of the work being carried out by responsible Electrical Contractors, who will provide wiring ample for the full capacity of the 'Outlets.'

SUGGESTED 'OUTLET' PROVISION

Showing the apparatus which might be in use.

These suggestions are in addition to one good Lighting Point in each room.

KITCHEN-SCULLERY:

3 'Outlets'

Water Heater, Fan, Iron, Washing Machine, Kettle (Refrigerator). *The wiring for the Cooker to be arranged for direct connection to the switch controlling the Cooker.*

LIVING ROOM:

2 'Outlets'

Fire, Vacuum Cleaner, Fan Kettle, Standard Lamp.

DINING ROOM (Parlour)

2 'Outlets'

Same as Living Room.

HALL:

1 'Outlet'

Fire, Vacuum Cleaner.

BEDROOMS:

2 'Outlets'

Fire, Vacuum Cleaner, Standard Lamp, Curling Tongs.

BATHROOM:

2 'Outlets'

Water Heater, Towel Rail, Fire.

LANDING:

1 'Outlet'

Vacuum Cleaner.

N.B.—Many Members also recommend Lighting Points at Front and Back Porches and in the Larder.

2.5   Electrical Association for Women, *Scientific House Management: The Fourth International Congress for the Scientific Organization of Work*, 1929

The Fourth International Congress for the Scientific Organisation of Work,

Source: Electrical Association for Women, 'Scientific Home Management: The Fourth International Congress for the Scientific Organization of Work, *The Electrical Age*, vol. 1, 1926–1930, pp. 555–559.

which was recently held in Paris, marks a forward step in the evolution of household science. Without doubt, after the industrial section, that devoted to the subjects of home management was the most interesting, detailed, scientific, and followed with greatest enthusiasm by its delegates.

Here is a brief picture of the household section in its three aspects— physical, intellectual and moral: an onlooker would have seen an animated group, attentive, interested, averaging 80 persons from day to day, and representing countries as widely different in location and race as France, Spain, Belgium, United States, Czechoslovakia, Poland, Italy, Holland, Bulgaria, Peru, Germany, Switzerland, Greece, Hungary and Egypt. It was remarked even by the men how closely the women kept to the strict schedules of their sessions. While the reports themselves revealed the scientific university training of the women addressing the Congress.

The outstanding ideas presented in the reports were as follows:

1 The American reports stressed chiefly the increasing development of comfort and convenience in the home: improved household equipment, and the education of women in household subjects; also the increasing aid given by industry to the solution of household problems.

2 The Belgian reports brought out the importance of adding the element of 'sport' to studies of household tasks followed with stopwatch and pedometer. They suggested adding elements of skill, endurance and speed—qualities of the new age—to the traditionally uninteresting household occupations. They cited as examples the tests made on children of from 13–18 years, who, by these methods, increased their rapidity and 'output' by 100 per cent.

3 The reports from France covered an exhaustive study of dust and its removal. As Mme Vèzes (President of *La Ligue D'Organisation Ménagère* of France) pointed out, the problem of dust removal is not confined to the housewife, but is rather a matter deserving municipal study and control. This view point—that many household problems must be solved and aided by the community—was further carried out in the reports of the following architects and engineers: MM Ziska, Gandillion, and Le Corbusier, as well as Paulet (from Peru, S. America). They substantially agreed that household operations must be more and more cared for by the municipality; and cited the case of community central heating, water supply, garbage and dust removal by means of inter-communicating disposal chutes; light, heat, and power (the 'big three' of the household), furnished at low cost by central electric stations—yes, even central refrigerating plants, for the more perfect control of house temperature. In short, they showed a universal feeling that there must be at one and the same time 'an increase of comfort in the home simultaneous with the suppression of manual drudgery.'

4 The report from Holland was in the hands of Mme E. J. Van Waveren-Resink, who, with Mme Muller-Lulofs, were recent delegates at the Newcastle Conference. Mme Van Waveren made the excellent and constructive suggestion that there be established *an international centre for the training of efficiency experts*. International leaders should prepare a correspondence course in these subjects: the standardisation of household tasks, the choice and operation of labour-saving appliances, the rationalisation of space and house-building, the application of business principles to the management of home finances. These courses would be available to young women who had already passed their training in traditional domestic

science; and these graduates would then receive a certificate as household efficiency experts, who could then advise any home manager on any problem of the home. Just as in the factory there are efficiency engineers, consulted by the manufacturer, so these women would form a new profession of 'household engineers' qualified to consult with the housewife—or the manufacturer—as to the efficient installation, method or practice which should be followed in modern home engineering or management.

5 The chief report from Germany was made by Mme Hildegarde Margis, President of the Housewives' League of Berlin. Her chief point was that the design and type of household appliance used in the home is of concern to both manufacturer and housewife alike. She urged that the Government establish standards of testing appliances, and issue a certificate or seal of approval on all household products before they are manufactured. She compared the pure food control against impure foods, and urged that a similar bureau of standards be set up by Government for the housewife's protection against inefficient and faulty appliances. An interesting report of the way efficiency is being taught in some German schools was given by Frau Silberkuhl-Schulte, of Berlin. She explained how the children were taught economy of effort by means of attaching string to different pieces of furniture, so as to arrange that equipment in the most step-saving way.

6 An intensive analysis of work in the home laundry was given by Mrs Christine Frederick, pioneer in the field of applying the scientific principles of motion-study to household tasks. Mrs Frederick, author of *Household Engineering* and other books on home management, and recalled as having given a series of lectures for the Electrical Association in England two years ago, brought with her a model Efficiency Laundry on a diminutive scale. This model, completely fitted with small-scale appliances of washer, ironing machine, etc., and complete electric lighting, served to visually illustrate the principles of step-saving and grouped equipment as applied to the washing of clothes. Mrs Frederick pointed out the great saving in time, effort and attention if an electrically operated washer was used, over a hand method. Electricity, said Mrs Frederick, permitted the use of an entirely new washing principle—and it was necessary to stress the education of the woman in the use of the machine. This education should be carried on by the manufacturer, as well as by the demonstrator of the appliance.

7 The delegate from Italy, Mme Diez-Gasca, sent in a report calling attention to the need for estimating the value of the housewife's own time in determining the cost of any household task. If women are expected to remain at home then they should receive compensation for their work as household managers. Both the delegates from Poland, Mme Szumlakowska, and the delegate from Check, Mme Pelantova, agreed in this view and gave numerous examples to corroborate it.

8 Last, but not least, much of the helpfulness of the Congress was due to the splendid direction of Mlle Paulette Bernege, the brilliant editor of *Mon-Chez-Moi*, and the founder of the French *Ligues d'Organisation Ménagère*. An admirable executive, Mlle Bernege, marshalled her delegates and ordered the presentation of the various papers with skill and helpful comment. In her talk before the men delegates of the convention, Mlle Bernege said in part: 'Gentlemen, it is apparent that you think of the amelioration of your own personal work; it is necessary that you think of the

improvement of the work of your factory operatives; but is it not equally important that you think of the work of your wives? You like to be surrounded by a happy family, and wives who are not too overburdened. Do as much for them as you have done for your workers, to the end that the house shall not be a nightmare, but a centre of happiness and peace— the organisation of household work from the point of view of modern scientific efficiency makes for the greatest human values—the health of all, the well-being of the wife and mother, and the harmonious progress of the family.'

2.6   Electrical Association for Women, *The EAW All-Electric House*, 1936

FOREWORD

Great strides have been made in the last few years in providing well planned, comfortable and pleasing dwellings for artisans, and again in the building of luxury houses and flats, but the production of houses for people of moderate income, who prefer to live in houses of better taste than the average villa, has been much less encouraging.

It is for this huge section of the community that this house has been designed.

In building it, the Electrical Association for Women fixed the cost at £1,000, and decided, in the best interests of housewives, that the house should be completely electric.

The chief aims have been:

To obtain reliable facts and figures.

To ascertain what size of house could be built with reliable materials, workmanship, equipment and good design, for £1,000.

To encourage Architects to specify, and Builders to include, as much electrical equipment as possible in the price of the house.

To observe how far Manufacturers could supply materials and equipment asked for by women at reasonable prices.

To have some ideas incorporated in the design which would be of assistance in the running of the smaller house.

To encourage women, by showing them an example, to ask that electrical equipment should be included in the price of their houses, and to stimulate them to take more interest in the general design, construction and wiring of their houses.

The Association hopes that these aims have been successfully achieved.

[. . .]

Electrical Equipment. The electrical equipment included in the price of the house is as follows:

   3 Directional Fires.

   6 Inset Fires.

 12 Feet Tubular Heating.

   3 Electric Clocks.

   1 Linen Cupboard Heater.

Source: Electrical Association for Women, *The EAW All-Electric House*, London: EAW, 1936. Pamphlet in the EAW Collection, Institution of Electrical Engineers.

1 Towel Rail.
1 Refrigerator (2 cubic feet).
1 Fan.

Hired Apparatus:
1 Water Heater, 1½ gallons.
1 Bath Water Heater, 15 gallons. (This also supplies two basins.)
1 Wash Boiler, 10 gallons.

Although the cooker hired by the local Electricity Department is not shown for the purposes of the exhibition, this would probably also be hired.

Other Electrical Apparatus Exhibited:

Hotlock Trolley.
Washing Machine with Wringer.
Ironing Machine, and Food Mixer.
Kettles.
Irons.
Coffee Percolator.
Hotplate.
Fan.
Vacuum Cleaners.
Floor Polisher.
Dusting Machine.
Clock.
[. . .]

Toaster.
Nightlight.
Poker.
Cigarette Lighters.
Reading and Standard
⌊ Lamps.
Small Food Beater.
Immersion Heater.
Bed Warmer.
Milk Warmer.
Hair Dryer.
Curling Tong Heater.

Special Features:
Long room for entertaining. Can be divided into two by fabric on a runner. Wooden uprights at each side are grooved to receive the curtain to prevent a gap.
Writing recess.
Built-in pelmets.
Warmed toy cupboard.
Small pylon at bottom of stairs to house telephone, telephone books and electric fire. Space alongside for a seat. Top of pylon large enough for writing comfortably.
Solidly constructed staircase balustrade.
No moulding anywhere to require dusting.
Glass and tiled window sills.
Rounded angles in plaster internal and external.
Service hatch in Dining Room, and return hatch for dirty dishes.
Dining Room window, which is of sliding, folding, type, opens right back to give unobstructed opening.
Lighting boxes each side of this special window to illuminate the curtains. This looks particularly well at the end of the long room when the curtain between the Living Room and Dining Room is drawn aside.
Built-in refrigerator.
Built-in kitchen fitments.
All flush doors.
The kitchen door, which can be made to occupy two positions, to shut off the hall while permitting direct access from the kitchen to

hall, and give direct access from the hall to the Dining Room.

A service hatch in the back door so that tradesmen can be answered without opening the door, and also a rack for two milk bottles outside.

Drying cupboard for wet washing.

Space for dustbin under cover.

Drawer in linen cupboard and some shelves sub-divided.

All cupboards lit automatically.

Built-in wardrobes.

Lighting box over front door serving three purposes: throws light down the door, illuminates the name and bell push, and lights the interior of the hall.

Sun bathing space on roof sheltered by walls.

Screen walls for privacy each side of the house.

It is appreciated that these items are all probably included in larger houses, but modern treatment shows how they can be introduced into the smaller house as well.

[. . .]

REMARKS

Selecting a normal suburban plot, and fixing the price of the house at £1,000, has naturally had a restrictive effect on the design.

It was thought better to aim at four bedrooms of smaller size rather than three larger. Built-in wardrobes and proper places or alcoves arranged for dressing tables, and having no fireplaces or chimney breasts, allow the greatest use to be made of available space, and it is the modern practice for bedrooms to contain only the minimum of furniture while having the maximum of air.

A larger kitchen or separate scullery for washing might be preferable, but this house is intended to be easily worked without a maid, and so the money has been spent in properly equipping the one kitchen.

The larder is small, but the dry goods usually accommodated in it have a special cupboard, and there is also the refrigerator to hold some of the food.

The drying cupboard removes the wet washing problem altogether from the kitchen.

The Hall, though small, is well arranged, and the large cupboard opening out of it and the ample cloakroom are improvements for the small house. The stairs, not being visible from the front door, is also a desirable feature.

Built-in fitments throughout the house save a good deal in furnishing and add to available space in rooms, and no coal boxes, fireirons or hearthrugs need be purchased. The wooden pelmets save material on all curtains.

A very ample supply of lighting and power points has been provided all over the house for convenience in using electrical equipment and to avoid long trailing flexes.

The cost of building materials and wages having advanced since the building was started, the garden work, which was originally intended to be included in the cost of the house, is now an extra.

COSTS

House and equipment, £1,000.

The land is owned by the builder and can be bought or taken on a ground rent by arrangement with him.

The standing charge for electricity, based on rateable value, will be approximately £6 per annum, the cost of electricity under the Bristol

Domestic Tariff being ½d. per unit.

Hiring charges:

Bath Water Heater	7/6 per quarter.
Small Water Heater	2/6 per quarter.
Washboiler	2/6 per quarter.
Cooker	3/- to 5/- per quarter according to size.

It is difficult to calculate the consumption of electricity as it is dependent upon the number and mode of living of the occupiers, but with a **very liberal use** for an average family it should not be more than £30 per annum, including the standing charge and all hiring charges.

It must be remembered that this cost covers the running of the whole house, including cleaning, washing, ironing, refrigeration, cooking, wireless, and hot water. There will be no bills for any alternative methods of heating, water heating or cooking, coal, wood, matches, and cleaning materials for grates. Ice will be always at hand, rooms will have added usefulness because immediate heat is available in all of them, costs for decorations and cleaning of materials will be appreciably lower, and there is a great factor of time saving. It will also be noted that fires can be used in all rooms instead of being restricted to the usual three rooms in a house of this type.

2.7   C. F. Scott, *What Was the Niagara Central Station Plan?* 1938

CURRENT PRACTICE – 1890   vs.   **NIAGARA PLAN – 1895**

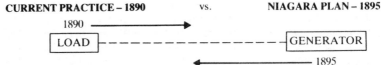

Current Practice began with <u>LOAD</u>– lamps or motors, and adopted different kinds of current for different lamps and motors.

Different circuits carried the appropriate kinds of current to the several kinds of loads: some economically limited to short distances.

Each kind of current requires special type of generator. Each generator limited to its own kind of service.

Most generators inherently limited in size. In some cases similar generators did not run in parallel, but each supplied independent circuits.

Product of individual inventors, intent on a particular service.

Product of the '80s when cut-and-try methods prevailed.

Product of cut-and-try search for a way to operate certain lamps or motors.

Niagara Plan began with <u>GENERATION</u>. A single kind of current for all purposes.

A universal transmission system suited to very great distances supplies current to substations to be transformed as desired for different types of load.

Generators may be operated in parallel as a unit.

A generator may be made of very great size, usually limited by the capacity of the power source which drives it.

Product of a broad scientific study of wide scope.

Product of the '90s based on prior empirical experience and on the rapidly developing scientific knowledge and engineering methods of design and operation.

Product of comprehensive research and coordination of old and new in a single Comprehensive System.

Source: C.F. Scott, 'What was the Niagara Central Station Plan?', typewritten memorandum, 5 April 1938; Niagara Archives, Georges Arents Research Library, Syracuse, NY. Reproduced in Thomas P. Hughes, *Networks of Power: Electrification in Western Society, 1880–1930*, Baltimore. The Johns Hopkins University Press, 1983, p. 123.

3

Materials

3.1 L. H. Baekeland, *The Synthesis, Constitution and Uses of Bakelite, 1909*

[. . .] I take about equal amounts of phenol and formaldehyde and I add a small amount of an alkaline condensing agent to it. If necessary I heat. The mixture separates in two layers, a supernatant aqueous solution and a lower liquid which is the initial condensation product. I obtain thus at will either a thin liquid called *Thin* A or a more viscous mass, *Viscous* A or a *Pasty* A, or even if the reaction be carried far enough, a *Solid* A.

Either one of these four substances are my starting materials, and I will show you now how they can be used for my purposes.

If I pour some of this A into a receptacle and simply heat it above 100° C., without any precaution, I obtain a porous spongy mass. But bearing in mind what I said previously about dissociation, I learned to avoid this, simply by opposing an external pressure so as to counteract the tension of dissociation. With this purpose in view, I carry out my heating under suitably raised pressure, and the result is totally different.

This may be accomplished in several ways, but is done ordinarily in an apparatus called a Bakeliser. Such an apparatus consists mainly of an interior chamber in which air can be pumped so as to bring its pressure to 50 or better 100lbs. per square inch. This chamber can be heated externally or internally by means of a steam jacket or steam coils to temperatures as high as 160° C. or considerably higher, so that the heated object during the process of Bakelising may remain steady under suitable pressure which will avoid porosity or blistering of the mass.

For instance, if I pour liquid A into a test-tube, and if I heat in a Bakeliser at say 160—180° C., the liquid will change rapidly into a solid mass of C that will take exactly the shape of its container; under special conditions it may affect the form of a transparent hard stick of Bakelite. It is perfectly insoluble, infusible, and unaffected by almost all chemicals, an excellent insulator for heat and electricity, and has a specific gravity of about 1.25.

It is very hard, cannot be scratched with the finger nail; in this respect it is far superior to shellac and even to hard rubber. It misses one great quality of hard rubber and celluloid, it is not so elastic nor flexible. Lack of flexibility is the most serious drawback of Bakelite. As an insulator, and

Source: L. H. Baekeland, 'The synthesis, constitution and uses of Bakelite', *Chemical News*, 23 April 1909, pp. 215, 220.

for any purposes where it has to resist heat, friction, dampness, steam or chemicals it is far superior to hard rubber, casein, celluloid, shellac, and in fact all plastics. In price also it can splendidly compete with all these.

Instead of pouring liquid A into a glass tube or mould I may simply dip an object into it or coat it by means of a brush. If I take a piece of wood, and afterwards put it into a Bakliser for an hour or so, I am able to provide it rapidly with a hard brilliant coat of Bakelite, superior to any varnish and even better than the most expensive Japanese lacquer. A piece of wood thus treated can be boiled in water for hours without impairing its gloss in the slightest way. I can dip it in alcohol or other solvents, or in chemical solutions, and yet not mar the beautiful brilliant finish of its surface. But I can do better, I may prepare an A, much more liquid than this one, and which has great penetrating power, and I may soak cheap, porous soft wood in it, until the fibres have absorbed as much liquid as possible, then transfer the impregnated wood to the Bakeliser and let the synthesis take place in and around the fibres of the wood. The result is a very hard wood, as hard as mahogany or ebony, of which the tensile, and more specially the crushing strength, has been considerably increased and which can stand dilute acids or water or steam; henceforth it is proof against dry rot. I might go further and spend a full evening on this subject alone, and tell you how we are now bringing about some unexpected possibilities in the manufacture of furniture and the wood-working industry in general. But I intend to devote a special evening to this subject and show you then how with cheap soft wood we are able to accomplish results which never have been obtained even with the most expensive hard wood.

In the same way I have succeeded in impregnating cheap ordinary cardboard or pulp-board and changing it into a hard resisting polished material that can be carved, turned, and brought into many shapes. I might take up much more of your time by simply enumerating to you the applications of this impregnation method with wood, paper, pulp, asbestos, and other fibrous and cellular materials; how it can be applied for fastening the bristles of shaving brushes, paint brushes, tooth brushes; how it can be used to coat metallic surfaces with a hard resisting protecting material; how it may ultimately supplant tin in canning processes; but I have no doubt that your imagination will easily supply you a list of possible technical uses even if I defer this subject for some other occasion. [. . .] As to Bakelite itself, you will readily understand that it makes a substance far superior to amber for pipe stems and similar articles. It is not so flexible as celluloid, but it is more durable, stands heat, does not smell, does not catch fire, and at the same time is less expensive.

It makes excellent billiard balls, of which the elasticity is very close to that of ivory; in short it can be used for similar purposes like knobs, buttons, knife handles, for which plastics are generally used. But its use for such fancy articles has not much appealed to my efforts as long as there are so many more important applications for engineering purposes.

Bakelite also acts as an excellent binder for all inert filling materials. This means that it can be compounded with sawdust, wood pulp, asbestos, colouring materials, in fact with almost anything the use of which is warranted for special purposes. I cannot better illustrate this than by telling you that here you have before you a grindstone made of Bakelite, and on the other hand, a self-lubricating bearing which has been run dry for nine hours at 1800 revs. per minute without objectionable heating and without injuring the quickly revolving shaft.

If I mix Bakelite with fine sand or slate dust I can make a paste of it which can be applied like a dough to the inside of metallic pipes or containers, or pumps, and after Bakelising this gives an acid-proof lining very useful in chemical engineering.

Valve seats which are unaffected by steam, steam-packing that resists steam and chemicals, have been produced in a similar way.

Phonograph records have been made with it, and the fact that Bakelite is harder than rubber, shellac, or kindred substances indicates advantageous possibilities in that direction.

For the electrical industry Bakelite has already begun to do some useful work. There too its possible applications are numerous. Armatures or fields of dynamos and motors, instead of being varnished with ordinary resinous varnishes, can simply be impregnated with A, then put into a Bakeliser, and everything transformed into a solid infusible insulating mass; ultimately this may enable us to increase the overload in motors and dynamos by eliminating the possibility of the melting or softening of such insulating varnishes as have been used until now. But the subject of dynamos and motor construction is only at its very modest beginnings, and I prefer to mention to you what has been already achieved in the line of moulded insulators of which you will find here several very interesting samples.

This brings me to the subject of moulding Bakelite.

For all plastics like rubber, celluloid, resins, &c., the moulding problem is a very important one. Several substances which otherwise might be very valuable are useless now because they cannot economically be moulded. The great success of celluloid has mainly been due to the fact that it can easily be moulded. Nitrated cellulose alone is far superior in chemical qualities to celluloid, but until Hyatt's discovery, it could only be given a shape by an evaporation process, and its applications were very limited. The addition of camphor and a small amount of solvent to cellulose nitrate was a master-stroke, because it allowed quick and economic moulding.

In the same way white sand or silica would be an ideal substance for a good many purposes, could it be easily compressed or moulded into shape and into a homogeneous mass. But it *cannot*; and therefore remains worthless. And that is the main difference between a *plastic* and a *non-plastic*. [. . .]

I have already shown you how I am able to mould and harden quickly by pouring liquid A into a mould and heating it in a Bakeliser. But even that method is much too slow for most purposes. Furthermore, moulds cost money; any rubber or celluloid manufacturer will tell you that the item of moulds represents a big portion of the cost of his plant. If an order for 10,000 pieces has to be delivered, and it takes an hour for moulding, it will require between three and four years to fill this order with one mould, and if the mould cost 100 dols. it will require 5000 dols. for moulds alone if the order has to be finished within twenty days. For that very reason I have devised my moulding methods so as to use the moulds only during the very minimum of time. I have succeeded in doing so in several ways. [. . .]

3.2   Hugh White, *Building Construction*, 1928

[. . .] The contrast between the older buildings and the newer is usually only brought into relief when something of a sensational nature, like a sweeping revision of housing regulations or a disastrous fire, brings their modern inadequacy to mind. The newer buildings of course command our closest attention as they attain greater and greater heights, although the engineering development which has made the skyscraper possible is only one phase of building progress. In comfort, sanitation, light and air, and a thousand other conveniences which are the products of the past hundred years, the present building, of however unimposing a type architecturally, far excels the most luxurious palace that was ever built for a Doge of Venice or a Louis XIV.

Nor are buildings, even of comparable grandeur, built to-day with the sacrifice of the health and lives of the workers, or by the lash-driven toil of men from the prisons, or the galleys, or even from poor districts where the scarcity of the bare necessities of life made necessary their unregarded labor at long hours and poor pay. All these conditions have been finally and definitely abolished only within the past hundred years, and they form a rare contrast with conditions of to-day, when the workers, under vastly different conditions, are building modern structures which they themselves, or others earning about the same wage, will later occupy as homes. The newer apartment houses for the working classes are excellently planned buildings, the difference between which and the more expensive apartment dwellings probably only consists of room size, and the extent of the ornamentation and the less important luxurious appointments. [. . .]

Another result of the social reform, so far as actual building construction is concerned, is made manifest in the numerous safeguards which surround the worker to-day against the hazards of accident. Where, formerly, low wages drove the worker to his task in a position which might, through the careless attitude of an indifferent employer, court injury or death, he is now equipped with the most highly improved devices for preventing accident to himself or his coworker. Openings are temporarily barricaded; scaffolds are supported by steel cables and special winches, and the factor of safety in any machine which might otherwise imperil the workers is made extremely high.

If in spite of these safeguards the worker is injured, compensation insurance or legal redress through laws which have been framed to protect him will, so far as money and skilled medical attendance are able, ameliorate any distress that he or his family might suffer.

This close connection between the social and political aspects of building and of the everyday life of the people would almost seem to make the history of building the history of its age. [. . .]

With this same concentration of great masses of people at their daily work came also the first building on any large scale of the community dwelling, the apartment house of our cities, the flat building of the English. The first of these were probably far worse, from the hygienic standpoint,

Source: Hugh White, 'Building construction', in Frederick W. Wile, *A Century of Individual Progress*, New York: Doubleday Doran, 1928, pp. 399–412.

than the old-law tenements which are the central feature of municipal reform discussions in New York at the present time; we are familiar, from descriptions in Dickens' works, with the slum conditions in London of a hundred years ago. It seems a far cry to the modern American apartment buildings, which good design, steel, modern invention and far-seeing legislation are making veritable palaces of habitation, beside which even the nobles' castles of a former time seem pitiably barren and comfortless.

Structurally, however, the modern building of steel is much less than a hundred years old. The first modern steel-framed building was erected in Chicago in 1885 by W. J. B. Jenney and George A. Fuller; it was the precursor of thousands which have followed this new method of construction, which, in conjunction with the development of the elevator, in one motion swept away the structural limitations and the restrictions which had hampered even the genius of the classical architects, but which fell before the vital demands of a new era of thought and action in building.

The main structural limitation theretofore had been the load-carrying capacity, and the thickness of the enclosing walls, which features fixed the economical height to which a structure could be built, but which condition has been eliminated almost universally by the skeleton steel frame in large building work. Curtain or panel walls are now supported at each story of these immense buildings on spandrel beams, these in turn being anchored to the tough, immensely strong steel columns which are the bone structure of the building.

In the place of the heavy and set feeling which the old and squat buildings engendered in the mind of the onlooker, we now have, from an architectural standpoint, the flexible vital upspringing of the steel frame in its light sheath of masonry.

The scientifically calculated loads of the building are now transmitted down through a system of steel columns to the ground footings, instead of an almost total concentration of the weight on the exterior walls.
[. . .]

Concurrently with the perfection of skeleton steel construction progressed the development of the second great factor in modern building. This is the reinforced concrete building, which is literally poured into place as liquid concrete, between temporary wooden or steel forms. Every inherent advantage of the flexibility of that material is thus utilized to high degree, and even the detail ornamentation of such a building is usually of concrete, although many of the individual pieces may be cast at a distance from the job and set into place during the finishing operations on the structure.

The reinforcement of the flat-slab type of building is by steel, usually in the form of rods and mesh fabric. Twisted and 'deformed' steel bars go into the heavier supporting pieces of poured masonry, while the floors are constructed by casting comparatively thin slabs of concrete over and around areas of steel mesh and smaller rods.

The chief use of flat slab construction is in buildings where great height is not the desired end. Factories, warehouses, piers and like structures are more and more being built by this method as one of the most economical and one of the most effective means of providing a fireproof and permanent building. The method has, too, a beauty of its own, as there is a certain moldable, flexible quality about poured concrete above and beyond its perfect adaptability to architectural creations derived from the classic conceptions which governed masonry construction.

The steel-framed and the flat slab building, then, are the two chief contributions of modern building to industrial progress to-day. Their significance is wide and deep, their potentialities boundless.

To industry it has meant that steel and concrete, both materials having the quality of adaptability in high degree, can follow closely and supply the always changing requirements of new mechanical conditions in manufacture. [. . .]

Steel has brought a great new age for architecture, just as the arch, the pendentive, or the flying buttress brought new forms and new conceptions into being. But modern steel buildings have a quality more inspiring to architecture than these old inventions with their limited scope. Steel has flexibility and life, and the buildings which it frames seem to have a fine exaltation in their cloud-piercing reach.

They express definitely the activity which goes on within their walls: the darting movements of the elevators within is seemingly caught in the pulsating vertical rhythm of tier on tier of windows.

The details of the science of mechanical equipment which was forced to keep pace with the growing demands of the vaulting building heights are now in themselves a vast and complex study. Supplying the modern skyscraper with the industrial and personal necessities of its thousands of inhabitants is a scientific undertaking, and calls for the highest exercise of a group of related talents.

The modern elevator, as an instance, is, in its finished and perfected state, a complex and highly developed mechanism which employs abstruse electrical and magnetic principles in its design. Speeds of 500 to 700 and more feet per minute are now common, while such features as the micro-drive, the unit multi-voltage control and the signal control have given us the efficiency and smoothness in operation which have come to be accepted as commonplace, although they would have been regarded with awe but twenty years ago.

The electrical and sanitary requirements of the modern large building, often of a capacity which would equal that of a fair-sized town, are also complete studies in themselves. Particularly so when it is remembered that the population of this building, unlike that of the city or town, moves in and moves out again eight hours later with clock-like regularity.

Provision must be made for supplying these people on every floor with thousands of gallons of hot and cold water, with electricity for lighting and operating the numberless household and office appliances, for insuring that there will be individual telephone connections, and for changing partitions and other stable fixtures as old tenants move out and new ones move in.

The first standpipe system is rigidly defined in its requirements by the codes of most municipalities and must form in itself a complete fire-fighting system, the capacity of which can be increased tenfold by connections made with the fire apparatus of the city.

Vacuum cleaner lines must penetrate the entire building and be connected with an elaborate suction apparatus in the basement. [. . .]

Engineer and architect must closely cooperate to-day because of the almost wholly technical nature of modern building methods. Each phase of construction, from the planning for excavation to the decision upon roofing types, calls for the combination of engineering and architectural faculties as represented in highly trained individuals, and specialization is increasing year by year the variety and the number of the experts involved in a large operation. [. . .]

It is interesting to know that architecture now recognizes no height limitation excepting that of economic construction. The economic point in construction is that which is passed when the cross-sectional area of the lower floors of the building is disproportionately occupied by service and mechanical facilities required for the upper floors.

That if elevators, immense pipes, and the necessary electrical, telephone. vacuum cleaning and other conduits occupy a great deal of the valuable space on the first floors, the advantages of the higher spaces are to a great extent nullified and the very immensity of the building may cause it to fail in bringing a suitable return on the investment.

Nevertheless, even greater buildings are being planned, and now there are several great underlying influences being taken into account in their planning. The city authorities now realize the immense demands placed upon transit facilities, upon water supply and upon street traffic by these great buildings, and future city plans will provide for their requirements. Changing conditions of transport, including the coming of air travel, will doubtless make necessary extensive changes in buildings, including provision for accommodating planes or dirigibles on roof tops.

But the flexibility of steel and concrete, their adaptability to any and all conditions, will probably go far to simplify these changes when they do come, and will make possible architectural marvels beside which even the immense buildings of our own age will seem small and inadequate.

Architecture itself must inevitably gain in the vitality which close association with the active life of a people means. There can be no stagnation in building when industrial trends bring it into ever closer relation to the lives of millions who inhabit large buildings or who earn their livelihood in them. This intimate connection between building and the lives of all the people is always tending to increase with greater concentration in cities and the social and industrial trends going forward with this concentration.

Ascetic and austere as may be the underlying art forms which control the design of the modern building it is now produced for a public need, with much closer relation to a people than the motives which inspired the great architectures of former times, so largely ecclesiastical as they were.

The personalities of the new era in building are a roster of names which include the greatest talent which the new world has produced. Engineers, architects and great financiers are those who have given us new forms and new principles to conjure with in seeking to lift the veil of future building.

3.3   John Hodge, *Workman's Cottage to Windsor Castle*, 1931

[. . .]

The Contract System

Messrs David Colville and Sons, Motherwell, in their early days, were a private company and were one of the very few firms who ran their melting department by the direct employment of all the workmen requisite and under their own management. The majority of the other firms let their melting shops out to contract, the contractor being responsible for the payment of the workmen's wages; they usually received a contract on an overhead basis at so much per ton. The result of this system was that the contractor and the men were always quarrelling about shortage of payments. In one case it would be that the men were being paid on less tonnage than they had produced, whereas the contractor would see to it that the full amount was paid by the firm; on repair work the same thing occurred; and it was a continuous fight to receive what was due. The same system was prevalent in all the works in England. It was many years after the initiation of the Trade Union that the contract system was abolished, and yet we were continuously pointing out to the heads of the various firms the inequity of the system and the continuous friction caused by it.

The building of the union

It must be remembered that the Siemens Steel Trade was practically a new industry. The men engaged in it represented all kinds and conditions of craftsmen, such as carpenters, joiners, blacksmiths, iron founders, ex-policemen, miners, and a very few old puddlers; all flocking to the steel works because of the glamour of the alleged big money which was being earned. It is interesting to note that although steel was ousting iron, no puddler would believe it, their minds being inflexibly made up that steel would very soon prove a failure and iron once again reassert itself. Therein lies the reason why so very few puddlers entered the steel trade. As indicated, however, the wages and conditions, year by year, were being continuously depressed as a result of this bad contract system, and the keenness of the contractors for big earnings, plus, of course, the desire of limited companies for big dividends. During these early days there was no Union. Consequently there was no sustained unity of purpose amongst the workmen in combating these continual reductions. I cannot recall a single strike which took place in those early days where the men fought successfully against a reduction in rates of pay, and if they fought, defeat was invariably their portion.

As already stated, out of the strike at Motherwell, when the men were subjected to the loss of one man out of three and 20 per cent reduction in their wage rates, there arose the determination that at last the time had come when an effort should be made to form a Union, and such was done. [. . .]

Source: John Hodge, *Workman's Cottage to Windsor Castle*, Marston: Sampson, Low, 1931, pp. 40–86.

Planning our programme

It must be remembered that up to this point, and for some months afterwards, I was still the unpaid general secretary of the organisation. I had visions and dreams of building a structure of organisation for the trade which would eliminate strikes and lock-outs by endeavouring to lay down foundations of good-will and honest dealing. My personal experience of strikes—four in about two years' time—made me loathe them. In order to perfect myself for the task I had undertaken, I made a study of how Parliament had dealt with labour during the past two centuries. I studied up the institution of conciliation boards in the iron trade, the founder of which was the late Mr David Dale, who was the chairman of the Consett Iron Company in the 'sixties of the last century. During the few short years I had been working in the steel trade, a participant or, should I say, a sufferer from strikes or lock-outs—call them what you may—I had a horror of them; their evil effects upon the character of the workmen and the sufferings of their wives and children, made an impression upon me which could not be eradicated.

My personal vow, therefore, was that everything that was humanly possible that one man could do to prevent such happenings, would be done. One curious thought struck me; that workmen did not like war between nations, realising that it would be the common people who suffered most from such events. The Trades Congress at times discussed the necessity of arbitration between nations. Conflicts between capital and labour may not have resulted in as much suffering as from wars between nations; but if arbitration in one case was right, it could not be wrong in the other. I went steadily forward along the pathway which I had planned, but it was only in 1904 that the fruition of these plans became a reality, when with the employers in Scotland and England we entered into a Sliding Scale Agreement for the regulation of wages of an automatic character.

Freedom from strikes

The paragraph headed 'Messrs David Colville and Sons meet their "Waterloo"' was the last strike—probably lock-out would be a more correct name—that the Steel Smelters' Union had until the catastrophe of the General Strike of 1926; that is, so far as organised workmen and organised employers were concerned. We were, however, in agreement with the employers that no outside employers should receive better terms than they gave to us. We have had with the organised employers usually through misunderstandings what one might describe as trifling stoppages, while the outside employer was compelled to accept the wages and conditions recognised between the employers and ourselves or face a stoppage. There can be no doubt that two organised bodies engaged in collective bargaining is the best method of regulating the wages and conditions of labour between the two sides. [. . .]

Charging machinery

The charging of the furnaces was all done by hand until the early part of the present century—hard, hot, heavy work and particularly exhausting when the weather was hot or sultry. The first firm to introduce the charging machine was the Parkgate Iron and Steel Company, Rotherham, the late Colonel Stoddart at that time being the managing director. When the new machine was almost ready for starting, the Colonel asked me to meet him for the purpose of discussing a reduction in the tonnage rates commensurate with the value of the machine. On the furnaces at Parkgate the number of men employed was four on each shift of twelve hours, and the Colonel desired that two men should be taken away from each group of four. One of these men was on datal rates and was paid by the firm; the third hand was a tonnage man. I suggested to the Colonel that instead of talking of the number of men who should be taken off each furnace, the wiser and the better plan was to make an arrangement as to what was the actual value of the machine and never mind the number of the men. Once we had agreed upon the value of the machine, the number of men would present no difficulty. My reasons for so doing were—and this appealed to the Colonel—that if you took away two men out of the group of four, per furnace, whenever any mishap occurred you would be shorthanded.

I gave as an example the furnace bottom getting into a bad state; frequently the work of putting it right was very hard work, heavy and hot; with a shortage of two men it would take much longer to do, and there would be a loss of output and on cost charges would materially increase. Let us, therefore, fix the value of the machine and leave the question of the number of men alone. Previous to my meeting the Colonel, I had explained to the workmen what my ideas were; fix the value of the machine and it would be better to distribute the tonnage wages rate which was left between three men than between two. To this they agreed and ultimately the Colonel saw the wisdom of my proposal. The value of the machine was fixed at $6\frac{1}{2}d$. per ton; the three men on tonnage rate remained, the rate so left being proportionately allocated between them. The datal man was kept on, so that the introduction of the machine, while the firm got its value, did not displace any of the workmen. This arrangement became effective wherever the machine was introduced, in England or in Scotland. [. . .]

The contract system

Reference has already been made to the contract system in melting shops. It was also prevalent in the rolling mill department. In fact, in 1886 it would have been a matter of extreme difficulty to get a single rolling mill, no matter what its nature or description, which was not operated by contract. In the sheet mill trade, for instance, one man might have a contract for one, two or more mills of both shifts, so that while he was sleeping comfortably in bed, the workmen were doing just as good work without his supervision as if he had been present, but the wages they received were miserable as compared with what they would have obtained if every man had been working on a piecework rate and had been paid direct by the firm.

In those early days (1888) my mind was made up that that system must be abolished. The consequence was that I took every possible opportunity of arguing with managers as to the evils of the system, impressing upon them that by splitting up a contract, and placing each man on a piecework rate, they would get greater outputs with equal efficiency and a more contented body of men. All this had some effect, at any rate, upon the mind of Mr Riley, who was always ready to listen to any workman who submitted any proposals for alterations in furnaces, which would make for their better working or any device which was likely to expedite outputs or lessen labour. As a matter of fact, Mr Riley was a great pioneer, so far as the Siemens steel trade was concerned, and always had a ready ear for suggestions. Other trades have suffered from this contract system in a somewhat similar way to the picture I have drawn, as instance in coal mining, you had it in what was known as the 'Butty System.'

Mr Riley and the contract system

Mr Riley transformed one of his plate mills by installing a new and more powerful engine equipped with an automatic reversing gear. New furnaces were also erected, his object being greater output, and to achieve this result, they were equipped with numerous labour-saving devices. The contractor for the operating of the mill was Mr Blackhurst, controlling both shifts of the mill. He started to operate the new mill under the conditions as respects his contract price per ton just as if no change had been made in the power or capacity of the mill. At the end of a fortnight's working, Mr Riley realised that in spite of all the machinery being more powerful, labour-saving devices and better and larger furnaces, his output was exactly the same as previously.

He therefore had Mr Blackhurst into his room, and I can imagine the kind of conversation that took place, and the final declaration of Mr Riley: 'Either you get 50 per cent, at least, increase in output, or I will find someone else who will.' Mr Blackhurst was a superior kind of man, as may be realised from the fact that year by year, as the Derby came round, he left the workmen to carry on while he was there for a week; Doncaster race week; Ayr races; weeks occasionally shooting; and the work still went on as efficiently as if he were there. At that time this contractor could not have been picking up less than £40 to £50 per week—I am inclined to think much more. Mr Blackhurst left the room still asserting that no man could do more than he had done.

After the expiry of another week or two, and no improvement, Mr Riley asked that I should meet him on a certain day for the purpose of discussing the problem. He poured out to me his tale of woe, the great capital expenditure of the new mill and furnaces, and the beggarly results. When he had finished, he said: 'Now, what do you think of it?' and my reply was: 'I think you are properly served.' This made him sit up. 'Why?' 'Have I not told you time and again about the evils of the contract system?'

Then I began to recite to Mr Riley how, year by year, his contractor went for a week to the Derby, and the work went on as efficiently in his absence as when he was present; his annual weeks at Doncaster; away shooting for a fortnight at a time; but the work went on efficiently. Then why not break the contract, put each man on tonnage rate, and then every man would have an interest in output, whereas working as they were on a datal rate, they had no interest in output. His next remark was, 'Can you take the men with you if I cancel the contract?' and my reply was, 'I can, but I want you to understand this, Mr Riley. I must have exactly the same rates and conditions that the contractor had.' He said, 'I agree.' I then said, 'Well, your word for me as to what that rate is would be sufficient, but, as I am dealing with the workmen, I want to see your books containing the rates, so that I can assure the men that I have seen the books and can certify that the figures are accurate.' And this was done.

Immediately after leaving Mr Riley, I summoned a special meeting of the whole of the men employed in the new plate mill in Blochairn, including the contractor. Previously I had worked out a rate for the contractor, who was now to become a roller, and also that he would require to work alternate weeks with the roller who would be appointed for one of the two shift turns. Instead of the contractor always being on days, coming when he liked and going when he chose, he must, under the new conditions, take his position as a rollerman. How he did fight against this proposal, but it was all to no effect. Minor alterations were made to the schedule that I had prepared and it was accepted by all the workmen, except the contractor, and immediately thereafter put into operation.

Instead of forty tons per shift, which the contractor gave, within six months that mill had more like an average of a hundred tons per shift. Those who know the trade will at once appreciate the enormous advantage this was, not only for the workmen employed, but for the firm, and the on-cost charges must have been materially reduced. The lead of the Steel Company of Scotland in this direction, and the publicity which was given to it, started movements in favour of the abolition of the contract system, with the result that firm after firm followed the lead. Consequently, for many years past the contract system in the iron and steel trade has been practically ended.

The contract system was not only an evil system as measured by the grinding of the workmen, but was bad management. This contract must have been exceedingly profitable to the contractor, as can be judged from the fact that some of them owned racehorses and ran them. How much better for the company and the workmen that a shop manager should be in charge, who would, as demonstrated in larger works than Dorman's was at that time, have done so with equal success, to say the least, than any contractor, and with a happier and more peaceful relationship between employer and workman.

3.4   H. V. Potter, *Plastics*, 1941

Plastics are a group of prepared materials whose properties of softening on heating with or without pressing render them extremely useful for shaping into finished articles which retain their shape on cooling. They are raw materials for many other industries. In fact plastics in some form enter into practically every known industry. [. . .]

The father of the modern technical plastic, namely, the synthetic resin or resinoids as we prefer to call them, was Dr L. H. Baekeland who

Source: H. V. Potter, 'Plastics', *Chemistry and Industry*, 8 March 1941, pp. 153–157.

in 1908 made the discovery of the phenolic resins and disclosed it in a series of worldwide patents. From that day there has been more intensive research on the resinoids and other plastics than I believe any other group of chemical products, with the result that while 35 years ago such a material was a novelty today there are hosts of different types of resinoids, many of which are available commercially. The phenolic, urea, and alkyd resinoid plastics differ from the earlier ones in their property of setting hard under applied heat known as thermo-setting and are peculiar to the alkyd, phenolic, and urea types whereas the earlier and other varieties remain permanently soft when heated and only harden on cooling. [. . .]

The phenolic resinoids revolutionized the industrial application of plastics on account of their property of softening initially when heated but by the continued application of heat they changed into a permanent hard product which would no longer soften on further heating. With most of the earlier plastics this process of softening and hardening was reversible by alternate heating and cooling so they are called 'thermo-plastic' whilst many of the resinoid plastics are irreversible and are known as 'thermo-setting.'

I have so far been referring to the materials in pure state, i.e., without fillers. As such they are usually transparent or translucent and in this state can be employed in the engineering industry to only a limited extent owing to their lack of strength and workability and their high cost. In most cases, therefore, fillers are added to the resinoids by mechanical processes so as to produce a uniform mass which can be readily worked or moulded and which in the finished form possesses higher mechanical and electrical strengths. Various fillers are employed according to requirements of the finished products and they include such materials as woodflour, mica, asbestos, graphite, paper, or fabrics or mixtures of some of these. Considerable resistance to mechanical shock can be imparted by the incorporation of suitable fillers consistent with a high degree of hardness and tensile strength especially when the fillers are fibrous in structure as distinct from granular. [. . .]

Building trade

[. . .]

Applications in the building trade are at present mostly for surfacing either as wall panels which may be natural wood finished or obtainable with an imitation wood finish or they may be finished in various bright colours. The materials comprise sheets and tubes which are usually of a laminated structure with a paper base. Panels of plastic materials are also used for such purposes as partitions, flush door surfaces, cupboard doors, shelves, window sills, skirtings and kicking boards, tops for tables, dressing tables, desks and other pieces of furniture, bath fronts, and washbasin splashboards.

An excellent example of a structural unit made from plastics is a fan complete with casing which is built up of individual mouldings assembled into one complete unit. The fan is mounted in window panes and serves the purpose of ventilating rooms of houses, offices, and factory workshops (in the last-mentioned case for exhausting corrosive fumes).

An interesting application in the building industry is a heat insulator for low temperatures made of cellulose acetate in crinkled sheet form. It is usually black in colour and the chief features in this application are low weight, high heat insulating value, non-porous, non-absorbent, all combined with ease of assembly. Such a plastic is also used for the transmission of light for such purposes as windows or structures for outdoor illuminations. Cellulose acetate, either sheet or tube, is successfully applied as an alternative to glass. Apart from its low weight it possesses the attractive feature of its inability to shatter and for this reason it is successfully applied for protecting lights in foodstuff factories, bakeries, and kitchens in order to avoid possible contamination of foodstuffs. In some cases the cellulose acetate is armoured with metal mesh. [. . .]

Shipbuilding

The shipbuilding trade employs plastic sheets for bathroom partitions, swimming pools, splashboards, cocktail bars, etc., as exemplified on such ships as the *S.S. Mauretania* and the *S.S. Queen Elizabeth* in addition to other famous ships on the Atlantic and Pacific services. An important mechanical application in shipbuilding of over two years' experience so far is the construction of a stern tube liner in a phenolic plastic instead of the usual lignum vitae on account of the much increased life of the former. The indications are that the plastic liner will have at least four times the life of the usual liner—perhaps a small but none the less important, successful application of a plastic material in the shipbuilding industry.

Railway engineering

The *Royal Scot* and other famous trains have plastic sheets fitted to a large extent in the corridors of the coaches as well as on the surfaces in the kitchen car such as wall panelling, cupboards, table and counter tops and in the cocktail lounge such as wall panelling, table tops, and window sills.

An example of a stressed member in railway engineering is the fish plate insulator on the track of electrified lines. This insulator is in continuous movement when trains pass over it and whilst it acts as an insulating medium it is also a supporting medium and it is found that very heavy forces operate in this case which require a material of great strength and durability for its construction. Phenolic plastic sheet is the basis of these fish-plate insulators. [. . .]

Silent and corrosion-resisting gears

These are used because of the desire to reduce noise generally in factories today. The phenolic plastic gear is mounted to mesh with a metal gear and not with another plastic gear. In terms of phons the reduction of noise is about 65% and this represents a difference between the noise of a quiet office and that of an average restaurant. In addition, these plastic gears in many cases outlast in life their metal counterpart whilst they are not affected by corrosive conditions such as acid liquids or gases. They can be used to transmit several hundred horse-power or a fraction of one horse-power, even so low as that in a small clock.

The careful attention given to the selection of suitable materials for mechanical purposes has to some extent led to the increasing use to which laminated materials are being employed for gear work. A considerable number of drives do not require gears made from exceptionally high tensile strength materials, neither in many cases is it economical to demand gears cut out to precision limits. In such cases laminated materials can be considered with advantage.

Briefly the characteristics claimed for these types of materials are that structurally they possess toughness unusually high for organic materials, also good wear resistance and a high degree of stability, being unaffected by atmospheric conditions, water, oils, steam, acid fumes and heat up to 250° F. They are quite vermin-proof and more silent in running than metal gears.

Generally two classes of phenolic laminated materials are employed for gear purposes. One is made with a coarse-grained fabric and is used for general purpose applications where the teeth are larger than 10 diametral pitch or the load to be transmitted is light; the other is a fine fabric material used for applications with small teeth, such as clockwork mechanisms require. This latter material is usually the stronger of the two and can be used to advantage in applications where a greater safety margin is required.

Laminated fabric materials can be used for the production of all types of gears and the prospective user should consider the manufacturers' recommendations with regard to power capacities, conditions of service, and machining in order to avoid unsuitable installations. [. . .]

Plywood industry

A great deal of experimental work has been done on the application of synthetic resins to the plywood industry. As bonding and impregnating mediums the main interest is the aeroplane industry. Both phenolic and urea resinoid materials are very extensively used today for this purpose in the form of: 1 thin tissue paper impregnated with the resin and dried, the paper merely acting as a carrier of the resin adhesive; 2 as liquid glues and cements applied by brush or spray.

In the first case phenolic resins are used and the adhesive sheet is produced in a similar manner to the ordinary coated laminated paper by passing the thin tissue through a bath of the liquid resin; it then passes into the drying tunnel travelling at such a speed as to deliver a dry thin sheet of varnished tissue at the other end. The tissue is shipped to the plywood manufacturer in rolls and he cuts off sufficient to cover the surface of the ply veneer that he wishes to press into a board, places the plys with the tissue cement in between into a hot veneering press, which liquifies the resin, causes it to polymerize and act as a waterproof bonding. [. . .]

The development of methyl methacrylate plastics as commercial products is one of the most interesting achievements of the industry. Methyl-methacrylate is a transparent thermoplastic resin. Following extensive research work full-scale manufacture of a resin of this type was commenced

in this country in 1934. The material is available in several forms:
1 as transparent sheets often referred to as synthetic glass,
2 as a moulding powder,
3 as cements and emulsions.
The properties of unfilled methacrylate mouldings are in general similar to those of transparent sheet.

The rapid expansion of the Air Force, together with the constantly improving performance of aircraft, called attention to the necessity for a light transparent material which could be readily shaped to form windscreens, cockpit covers, gun turrets, etc. This plastic has now been used in military and civil machines for nearly three years and is employed largely for such components as cockpit covers, windscreens, cabin windows, navigation windows, and numerous other transparent parts. Almost invisible cemented joints can be made, but riveted joints are commonly employed. Suitable clearances must be provided to allow for differences in the coefficients of expansion of transparent sheet and metal. It is of interest to note that synthetic resin windows transmit less noise than glass windows owing to the heavy damping properties of the resin. It may be said that the availability of suitable transparent synthetic resins has rendered possible certain forms of construction in aircraft which otherwise could not have been adopted.

Pile driving

A particularly interesting application which enabled a contractor to drive steel piling to a predetermined depth in very hard ground after he had given up in despair any hope of getting the piles within feet of their objective. Because of the nature of the ground driving direct on the end of the pile caused the end of the steel pile to mushroom. A cast steel helmet was employed. Normally, in these steel helmets a hard wood dolly is used, but the force required to move the piles was such that the wood dolly crushed and even burst into flames. A metal dolly was impracticable and it was not until some phenolic resinoid fabric material was tried that any hope of success was entertained. The size of the dolly was 20 in. by 9 in. by 3 in. thick, and came in contact with the rough cast bottom of the helmet. The hammer operating under a steam pressure of 80 lb. per sq. in. delivered 120 blows per minute, with a driving force of 15,800 ft. lb. per blow. Admittedly, the material did not stand up indefinitely, but one piece stood up to the equivalent of six hours' continuous working and then failed through being partly charred to destruction, owing no doubt to the heat set up by the excessive friction caused by the very uneven surface it was working on.

Rolling mill bearings

Laminated fabric materials have now been installed as bearings for the last few years. To give a general idea as to suitable applications for bearings made from resinoid products, it may be taken that where bronze, white metal, lignum vitae, etc. have been previously used these non-metal materials can receive equal consideration.

A British user has stated that with bronze bearings on a rolling mill it was the practice to renew them every 6 to 8 weeks, a set of resinoid laminated fabric bearings fitted to the same mill were not replaced until 20 months had elapsed, the conditions of service being the same throughout. Also it was estimated that the savings in power consumption were approximately 25% to 30%.

There appears to be a general opinion in favour of bearings made from resinoid materials both as regards life and power consumption. Other advantages which have been found are the ability to work satisfactorily with water only as lubricant, cleaner operation, reduction in maintenance charges, and cheaper operating costs, longer runs between shutdowns, cooler rolls, loading can be increased and heavier reductions made with consequent increase in production, also the use of grease is confined to short periods immediately prior to weekend stoppages.

Domestic appliances such as vacuum cleaners

Some vacuum cleaners available to the public today can almost be called plastic vacuum cleaners owing to the large number of components contained in them which are made of plastics, even including a large part of the body or housing for the moving parts, rollers, handle, motor armature, and other insulation, etc. A careful selection was made before a suitable plastic could be adopted. The fact is, however, that a suitable plastic exists from which exacting production is carried out on a large scale. It is employed to reduce weight.

Engineers' tools are now available in quantity fitted with moulded plastic handles each of which outlasts its wooden counterpart many times. Moreover, it is cleaner to handle and in cases where a good grip is required it is more efficient. Not only black, but other colours are used. The handle is moulded on to the shank of the tool as the last operation in production, as in the case of its wooden opposite number. The ribbing on the handle to give a good grip is produced during the moulding of the handle all in one operation.

Cams

Many machines such as those used for bootmaking, sewing, bottling, nailing, labelling, and stamping are cam-operated. The cam as a moving part must be well balanced and its weight must be reduced to a minimum consistent with strength in order to lower the inertia as much as possible. An increase in speed of operation is the general trend in many instances and this has been assisted in many cases by the introduction of cams moulded from special phenolic resinoid laminated material. Accuracy of profile and balance are obtained in the moulding operation. Lubrication can be assisted by the introduction of graphite and the many other properties including lightness in weight, shock resistance, wear resistance, and silent running, all play their part in helping the general speed-up of operation. Machines that could be run at 150 r.p.m. with metal cams have been speeded up to 450–500 r.p.m. by using plastic cams. [. . .]

Electrical engineering applications

These are more numerous than those of a purely mechanical interest owing to the good electrical insulating properties possessed by many plastics. Whereas rubber and vulcanized fibre, slate, and procelain were mostly used for electrical insulation at one time, the newer plastic materials have replaced these for many purposes. In addition the newer plastics are used not only in the form of sheets or slabs but are moulded to finished shape with moulded-in fittings in the one operation of forming the parts and therefore they have become widely adopted as finished structural components without the necessity of performing many costly machining operations on them before subsequently assembling to one whole unit. Two outstanding examples of the latter class of use are radio cabinets and telephones [. . .]

Conclusion

[. . .]

The plastics age is still in its early stages. New products are exemplified in the patent literature and new applications are of almost daily occurrence. When a new article is being considered for production, consultation with the supplier and those engaged in the industry should be considered. The choice of material to be used in the past has not needed much consideration. It usually fell into the class of metal, wood, glass, or ceramics. A fifth class—plastics—has attracted much attention recently and each day sees it claiming articles which previously would have been made in the other classes—not because of a prevailing fashion but because plastics were chosen after considering suitability, cost, ease of production, durability, and appearance. The heavy and many other industries are not so much concerned with appearance as with performance, although in many cases beauty accompanies the other properties automatically. [. . .]

3.5   G. Loasby, *Nylon Yarn*, 1944

[. . .] Fibres of animal and vegetable origin comprised the raw material for the manufacture of the greater part of the world's textiles for 5000 years and were adequate until the beginning of the present century. It is true that the production of rayons and synthetic fibres at the moment is still small in comparison with the annual cotton crop, as Sir Robert Pickard reminded us in his Mather Lecture, the relative proportions according to 'Rayon Organon,' 1942, being as follows:

	Million lb.	%
Cotton	12,700	68
Wool	2,440	13
Silk	100	1
Rayon	3,473	18

Of the rayon production, 58.3% is staple fibre

Source: G. Loasby, 'Nylon yarn', *Chemistry and Industry*, 5 August 1944, pp. 282–283.

The production of these newer fibres marks the beginning of an era in which the quality of textile raw materials is not so dependent on the internal economics of an individual producing country, nor does its annual production depend on prevailing temperatures, rainfall, and disease. This in itself is an enormous stride forward, but when to it is added the certainty that the new materials shall have previously designed textile characteristics and particular properties, one realizes the potentialities of the modern textile industry. [. . .] The established rayons and also nylon yarn have an evenness of denier which exceeds that of natural silk. Whereas the rayons had adequate strength for handling and processing into fabrics, they also possessed a price advantage over silk which brought the manufactured products within the reach of a mass of people for whom silk was too expensive.

[. . .] Nylon yarn in the form of continuous filaments is, of course, the first synthetic yarn in the true sense of the word, having been built up from simple products with a molecular weight of under 100 to complex molecules which lie between 5000 and 50,000, an increase of some 50 to 500 times. The historical development of nylon itself is mainly the record of the work of Carothers and his coworkers, who, from 1929 onwards, began to publish their researches on the building up of simple substances into complex bodies, showing how to imitate nature's achievements in synthesis.

When the molecules, or units from which these were constructed were sufficiently large and made up of long chains, Carothers found that the molten products could be drawn out into fibres which were strong, pliable, and resilient. All the compounds except one class, the polyamides, either decomposed or melted at too low a temperature, and the history of the nylons is chiefly of this class of substances.

Wallace Hume Carothers was born in Burlington, Iowa, in 1896. In 1932 he published a paper entitled 'Artificial Fibres from Synthetic Linear Condensation Superpolymers,' J.A.C.S. 54, 1579–87 (1932). This paper described the formation of a complex polymer thus: 'Continuous filaments can be obtained from the molten ester by touching it with a stirring rod and drawing the stirring rod away, as we might with a tube of glue.' These investigations led finally to the development of nylon, an all-synthetic fibre of very outstanding qualities. By 1939, researches had progressed so far that the first factory commenced operations on a commercial scale at Seaford, Delaware, in December of that year. By May, 1940, the production of full fashioned hosiery was at the rate of 64 million pairs per year. By 1941 the rate had passed the 100 million mark.

From the researches of Carothers to the present time, over one thousand varieties of nylon have been produced and some 400 patents taken out in the U.S.A. covering polyamides, polyesters, and so on. The present nylon is one of the first thousand. Continuous research in America, Canada, and this country is proceeding to improve the fibre in many respects. Two of the chief uses have been in parachutes and in aircraft tow-ropes. Since nylon is unaffected by rot and mildew, great use is also being made of it at the present time in stores for troops operating in the tropical jungles. [. . .]

Nylon yarn is specially characterized by a high tenacity and high elongation at the break—a combination of properties not yet found except in the naturally occurring silk, where it exists in a lower degree. The rayons can be spun with dry tenacities of the order of silk, but always at the expense of the elongation at the break becoming more or less brittle. Nylon yarns, on the other hand, possess an elasticity and resilience in a remarkable degree. This property makes them ideally suited for hosiery construction and other purposes. In addition to the special war-time uses and peace-time development in hosiery, nylon has found its way in filament form into shoe laces, slipper fabric or satins for dance shoes, foundation garments, brassieres, insect screens, etc. Millions of toothbrushes have been made from nylon bristle, or monofil, as distinct from multi-filament yarns. Hair brushes and industrial brushes are being made with these synthetic bristles, and such brushes have been made use of in textile printing. Fishing lines, leaders, casts, films, and surgical sutures are also being made. Nylon yarns and fabrics can be steam set under pressure, a property made use of in finishing hosiery, knitted goods, and fabrics. This process is known as preboarding. [. . .]

In dyeing properties, nylon yarn lies between wool and silk. Nylon fabrics and yarns can be dyed with a variety of dyestuffs, of which perhaps the Direct and Acetate types are the most popular. The most level dyeing are the Acetate types—Duranols, Solacets, Dispersols, and Acetamins. Direct types are apt to emphasize differences in dyeing affinity.

So far there has been little work done on peace-time development of nylon yarns for weaving fabrics—all the output of nylon both here and in the U.S.A. going into war-time products. Pre-war development was chiefly in the hosiery field. Continuous nylon yarns can be made into ninons, voiles, taffetas, shoe, corset, and other woven fabrics. They are also expected to find a market in knitted goods, warp-knitted fabrics, milanese, and lace. Monofils in limited quantities have been used for hosiery. After the war, nylon will not be confined to the hosiery trade, but as the supply is bound to be limited, development must be along those lines which show the most promise and in which the outstanding properties of nylon are used to the best advantage.

So far little has been said of staple fibre development. Nylon yarns are not yet available in the form of staple fibre, but in some ways spun yarns may be the most promising for the future. Fabrics from spun yarns are expected to possess increased warmth, drape and handle. These print well, as do fabrics of continuous filament yarn. It has been said that the speed of the production, processing, and weaving of nylon yarns in this country has been unequalled by that of any other fibre in its initial stage. This surely augurs well for its future, and we can look to a large development of nylon and nylon yarns as soon as the present crisis is over.

3.6   D.D.T. in General Use, 1946

During the war D.D.T. helped to control the insect vectors of such diseases as typhus, malaria, and dysentery. There was no need to emphasize the importance of this work, if the D.D.T. was properly applied with attendant success. Now D.D.T. is on sale to the public, and it will be used against a

Source: 'DDT in general use', *British Medical Journal*, 1946, II, pp. 203–204.

wider variety of insect pests and nuisances, but there may be failures and disappointments in store. For example, it is often thought that a few puffs of D.D.T. powder in the wardrobe will protect clothing from moths. But since D.D.T. has no action at a distance and no repellent effect, there is nothing to prevent moth grubs living safely under the lapel of a coat the top of which is covered with D.D.T. dust. A summary of the properties and methods of application of D.D.T. has been prepared by the Ministry of Supply.[1] It should be useful to everyone concerned with insect control.

Pure D.D.T. is a colourless crystalline solid (melting-point 108.5–109° C.), with a low vapour pressure (about 1.3×10^{-7} mm. Hg at 20° C.). It is chemically stable under most normal conditions, so that traces of the substance are very persistent. It is sparingly soluble in water—less than 0.2 parts per million—but soluble in varying degrees in organic solvents. It is least soluble in hydroxylic and polar solvents and most soluble in aromatic and chlorinated solvents. D.D.T. is lethal to a wide range of insects and other arthropods, but the lethal dose varies very considerably from one to another. The dose which will kill a mosquito or a fly is much less than that necessary to kill a flea or a louse, which again are more susceptible than bugs and cockroaches. To be effective the D.D.T., either dry or in solution, must come into direct contact with the cuticle. It is then absorbed and will poison the insect, but the effect is slow and may take from a few minutes to a few days, according to the size of the dose and the resistance of the insect.

$$Cl-\bigcirc-CH-\bigcirc-Cl$$
$$CCl_3$$

The most valuable property of D.D.T. is its persistence. Surfaces contaminated with it will poison insects which walk over them for many weeks after the original application. To bring D.D.T. into contact with harmful insects it is usually necessary to spread a small quantity of the insecticide over a relatively large area of clothing, walls, vegetation, or water. To achieve this the D.D.T. is diluted with a solid in the form of a powder or with a liquid to form a solution, emulsion, or suspension. It is possible to purchase commercial grade D.D.T. (about 70–80% pure) and make up a suitable preparation from it; but in most cases it is more satisfactory to obtain the finished product from a reputable firm. Such products should give the percentage of D.D.T. on the label.

In the form of a powder, diluted to 5 or 10% with any suitable inert mineral, D.D.T. can be applied against body vermin, such as lice and fleas, against pests of domestic animals, and against crickets, cockroaches, etc. The powder can be dispersed simply by shaking or by means of a 'dusting gun.' The most commonly used solvent vehicle is kerosene. A solution of 3–5% in this liquid produces a residual toxic film on walls and ceilings to combat such insects as bed-bugs and houseflies. A relatively coarse spray is most successful, since it is necessary to wet the wall; fine atomizing guns create mists which tend to float away. Emulsions and suspensions of D.D.T. in water will also form a residual film. Such preparations have been employed for the control of mosquito larvae and against head lice. It has been shown that D.D.T. can be incorporated in paint or distemper and retain some of its insecticidal power. Its toxicity to insects in this form, however, is much less than that of a superficial deposit, presumably because much of the D.D.T. is buried in the paint. Against flying insects it can be used in the form of a fine mist (aerosol) or a smoke. But these methods do not produce the lasting residual film which is the principal merit of D.D.T.

We have previously[2-3] reviewed the literature on the toxicity of D.D.T. to man and other mammals. Further results have appeared since then, but they have not substantially altered our conclusions. In insecticidal concentrations D.D.T. is practically harmless to man and domestic animals. The most likely dangers from careless use will probably follow accidental or intentional (suicidal) drinking of concentrated preparations, among which may be included the 5% solution in kerosene. There may also be some risk involved in repeated contamination of the skin by oily solutions. These hazards must be considered by pest-control operators, who will regularly handle concentrated D.D.T., but the general public will hardly be concerned with preparations of this type.

The earlier development of D.D.T. was to combat disease vectors and aid military hygiene. In the wider field of plant protection control operations are complicated by the danger of destroying beneficial insects, principally the parasites and predators which naturally keep pests in check. However, this risk has perhaps been overemphasized, since D.D.T. will only do more efficiently what other insecticides have done in the past. The agricultural entomologist is alive to the possibility and will proceed cautiously, keeping an eye on the effects of D.D.T. on the balance of population. The medical entomologist has no such anxiety, since the insect parasites of man are remarkably free from other parasites and predators.

Notes

1   Some Properties and Applications of D.D.T., H.M.S.O., 1946, 6d. net.
2   British Medical Journal, 1945, 1, 338.
3   Ibid., 1945, 2, 260.

4

Transport

4.1   E. A. Pratt, *A History of Inland Transport and Communication in England*, 1912

[. . .]

Thus far the railway certainly represents the survival of the fittest; and, curiously enough, although great improvements have been made in locomotive construction, in rails, in signalling, in carriage-building and in the various departments of railway working, no absolutely new principle has been developed since the Liverpool and Manchester Railway definitely established the last of the three fundamental principles on which railway construction and operation are really based: (1) that a greater load can be moved, by an equivalent power, in a wheeled vehicle on a pair of rails than in a similar vehicle on an ordinary road; (2) that flanged wheels and flat rails are preferable for fast traffic to flat wheels and flanged rails; and (3) that a railway train should be operated by a locomotive rather than by either animal power or a stationary engine.

It is true that, in regard to the last-mentioned of these three main principles, material changes have been brought about by the resort to electricity as a motive power; but this, after all, is an improvement in the means of rail transport rather than a complete change in the principle of transport itself; and, though electricity may supersede steam to a considerable extent, especially for suburban traffic, the resort to it is a reversal, in another form, to the earlier idea of motive power distributed from a fixed point, as originally represented by stationary engines, before the locomotive had established its superiority thereto.

In any case, the railway is still the railway, whatever the form of traction employed, and there is, after all, no such fundamental difference between an electric railway and a steam railway as there was between the railway and the canal, or between either railway waggon or canal barge and the carrier's cart travelling on ordinary roads. The question that really arises here is, not whether electricity is likely to supersede steam for long-distance as well as for short-distance rail traffic, but whether the railways themselves are likely to be superseded, sharing the same fate as that which they caused to fall on the stage-coach and, more or less, on the canal barge.

For the physical, economic and other considerations already presented, there is no reasonable ground for expecting much from the projected scheme of canal revival. When the country comes fully to realise (1) the natural unsuitability of England's undulatory surfaces for transport by

Source: E. A. Pratt, *A History of Inland Transport and Communication in England*, London, 1912, chapter 31.

artificial waterways; (2) the enormous cost which the carrying out of any general scheme of canal revival would involve; (3) the practical impossibility of canal-widening in the Birmingham and Black Country districts; and (4) the comparatively small proportion of traders in the United Kingdom who could hope to benefit from a scheme for which all alike might have to pay—it is hardly probable that public opinion will sanction the carrying out of a project at once so costly and so unsatisfactory in its prospective results.

Still less than in the case of canals would any attempt to improve the conditions of transport on rivers—serving even more limited districts, and having so many natural drawbacks and disadvantages—be likely to meet any general advantage or to foster any material competition with the railways.

Developments in regard to road transport are much more promising—or, from the point of view of the railways, much more to be feared—than any really practical revival of inland navigation.

Dealing, in this connection, first with personal travel, we find that the main competition with the railways proceeds from (1) omnibuses, motor or otherwise; (2) electric tramways, and (3) private motor-cars.

An omnibus, whether of the horse or of the motor type, is the equivalent of the carrier's van or of the old stage-coach in so far as it has the complete freedom of the roads. The electric tramway, while having to keep to a certain route, and involved in greater capital expenditure by reason of its need for rails, overhead wires and power stations, may, if owned by a local authority, still be materially aided, directly or indirectly, out of the local rates. Thus the omnibus and the electric tramway may both be able to transport passengers at lower fares than the railways, which, as regards the municipal tramways, may even be called on to pay, through increased taxation, towards the maintenance of their rivals.

In London itself the motor-omnibuses have undoubtedly abstracted a considerable amount of short-distance traffic from the Central London Railway, which, however, still has the advantage in regard to longer distance journeys.

That electric tramways and motor-omnibuses have also diverted a great deal of suburban passenger traffic from the trunk railways is beyond dispute. But here the companies are seeking to meet the position (1) by operating their own suburban lines by electricity, giving their passengers a quicker transport than they would get with tramways or motor-cars stopping frequently, or held up by traffic repeatedly, on the roads or streets; or (2) by offering to town workers greater facilities for removing from homes in the inner to homes in the outer suburbs, if not in the country proper or even on the coast itself—in other words, to such a distance that they would naturally be dependent on the railway and the business trains that are now run thereon from the places in question to meet their special convenience.[1]

Of these two developments the former has not yet been generally adopted, whereas the latter is in full activity, and, in combination with the heavier local taxation which is steadily driving people away from London boroughs, is helping to produce results of much interest and importance.

The population, not only of London, but of great towns in general, is undergoing a considerable redistribution. Land at greater distances from urban centres, and hitherto devoted only to agriculture or market gardens, is being utilised more and more for building purposes; the increasing values of land within the radius of these outer suburbs improves the position on urban markets of producers in rural centres whose lower rents may more than compensate for their slightly heavier cost of transport as compared with the suburban growers; the health of town workers taking to what are not merely suburban but country homes should improve. Social and domestic conditions generally are, to a certain extent, in a state of transition; while the trunk railways are getting back from their long-distance suburban traffic some—though not yet, perhaps, actually the whole—of the revenue they have lost on their short-distance traffic.

On the other hand, results are being brought about in the inner suburbs which are viewed with much uneasiness by the local authorities. The removal from the inner suburbs of considerable numbers of those who can afford to live further away from their business means (1) that population in the inner suburban circle is decreasing, or, alternatively, that a better-class population is giving place to a poorer-class one; (2) that much of the house property there is either standing empty or is fetching considerably lower rents; and (3) that the taxable capacity of the areas in question is declining, although the need for raising more by local taxation is to-day greater than ever.

Where the local authorities who are experiencing all these consequences of an interesting social change have themselves helped to bring them about by setting up municipal tramways to compete with the railways, thus, among other consequences, driving the latter to resort to measures of self-defence, they may find that attempts to change, if not to control, the operation of economic forces have their risks and perils; while the position for the authorities concerned will be even worse if their municipal tramway, in turn, should suffer materially from the competition of the motor-omnibus.

Private motor-cars may appear to have deprived the railways of a good deal of their passenger traffic, and they certainly constitute a most material and much-appreciated increase in the facilities now available for getting about the country. It must, however, be remembered that a very large proportion of the journeys taken in them would probably not be made at all if the motor-car did not exist, and if such journeys had to be made by train instead. The actual diversion of traffic from the railway only occurs when journeys which would otherwise be made by rail are made by motor, in preference. Here the railway certainly does lose.

Against the loss in the one direction in railway revenue, owing to the greater use of motor-cars, there can at least be set the constant growth in the taste for travel which the railway companies (partly, again, to make up for the competition in suburban traffic) have done their best to cultivate by means of abnormally low excursion or week-end fares based, as one leading railway officer put it to me, 'not on any idea of distance, but on the amount that the class of people catered for might be assumed to be willing to pay.'

The travel habit has thus undergone a greater expansion of late years than has ever before been known, so that a falling-off of railway traffic in some directions ought, sooner or later, to be compensated for by increases in others, if, indeed, that result has not already been attained.

The actual position in regard to passenger travel on the railways of the United Kingdom during the years 1901–10 is shown by the following figures, taken from the Board of Trade Railway Returns:

YEAR	PASSENGER JOURNEYS[2]	RECEIPTS FROM PASSENGERS £
1901	1,172,395,900	39,096,053
1902	1,188,219,269	39,622,725
1903	1,195,265,195	39,985,003
1904	1,198,773,720	40,065,746
1905	1,199,022,102	40,256,930
1906	1,240,347,132	41,204,982
1907	1,259,481,315	42,102,007
1908	1,278,115,488	42,615,812
1909	1,265,080,761	41,950,188
1910	1,306,728,583	43,247,345

These figures give evidence of, on the whole, a substantial advance in railway passenger journeys and receipts, notwithstanding all the competition of alternative facilities, and we may assume that although tramways, motor-cars, motor-omnibuses and even the latest new-comer, railless electric traction, may supplement and more or less compete with the railways, there is no suggestion that they are likely entirely to supplant them for passenger travel.

In the matter of goods transport in general, it is the fact that during the last ten or fifteen years, more especially, there has been an increasing tendency for the delivery of domestic supplies to suburban districts or towns within an ever-expanding radius of London and other leading cities to be effected by road, instead of by rail. The same has been the case in the distribution by wholesale houses of goods to suburban shopkeepers, and, also, in the reverse direction, in the sending of market-garden or other produce to central markets.

Where the railway companies have really created new suburban districts through the running of specially cheap workmen's trains, it may seem hard upon them that they should be deprived of the goods transport to which such districts give rise.

The fact must be recognised, however, that when the distances are within, say, a ten–, a fifteen– or even a twenty-mile radius, and when only small or comparatively small parcels or consignments are to be carried, the advantages in economical transport may well be in favour of the road vehicle rather than of the railway. The road vehicle can load up in the streets as it stands opposite the wholesale trader's warehouse; it pays nothing for the use of the road; it does not make any special contribution to the police funds in recognition of services rendered in the regulation of the traffic; nor is it taxed by the local authorities on the basis of the quantity of goods carried and the extent of the presumptive profits made; whereas the railway company must have a costly goods depôt, acquire land for their track, lay lines of rails, maintain an elaborate organisation to ensure safe working of the traffic, and submit to taxation by every local authority through whose district the goods carried may require to pass. There is, also, the further consideration, of which I have previously spoken, that in the case of short-distance journeys the cost of terminal services makes the rate per ton per mile appear much higher, in proportion, than when, while remaining at the same figure, it is spread over a substantially greater mileage.

While, with the increasing facilities for road transport, the railways must expect to lose more of their short-distance traffic, they should be able to retain their long-distance traffic, and more especially their long-distance traffic in bulk, commercial motors notwithstanding. Where commodities are carried either in considerable quantities or for considerable distances, and more particularly when both of these conditions prevail, transport by a locomotive, operating on rails, and conveying a heavy load with no very material increase in working expenses over the carrying of a light load, must needs be more economical than the distribution of a corresponding tonnage of goods among a collection of commercial motors, for conveyance by road under such conditions that each motor is operated as a separate and distinct unit.

The results, too, already brought about in the case of the suburban passenger traffic may, possibly, be so far repeated that railway companies deprived, also, of suburban goods traffic by the increasing competition of road conveyances, will show further enterprise in encouraging long-distance goods traffic to the same markets, or to the same towns. In this way they might seek to avoid, as far as practicable, any falling-off in their revenue at a time when taxation, wages, cost of materials and other working expenses all show a continuous upward tendency.

Should the policy here in question be adopted, market-gardeners, more especially, may find that, while they have effected a slight saving on their cost of transport by resorting to road conveyance, they will have to face increased competition from produce coming in larger quantities from long-distance growers who, with a lower cost of production, and also, with increased encouragement from the railways, might have advantages on urban markets fully equal to those of the short-distance grower located in the suburbs.

The whole question of the steadily increasing competition between road and rail has thus become one of special interest, at the present moment, alike for the trading, the motor and the railway interests.

That the use of motor-vehicles is destined to make even greater advance in the immediate future has already here been shown. Yet there are distinct limitations to its possibilities, although this fact is apt to be overlooked by motor enthusiasts, some of whom are, indeed, over-sanguine. One of them proclaims that 'the new locomotion' is 'designed to be the chief means of transit to be used by humanity at large,' and 'eventually will probably to a large extent supersede all others.' He further writes: 'Many of us will live to see railway companies in places pulling up their rails and making their tracks suitable for motor-car traffic, charging a toll for private vehicles and carrying the bulk of the traffic in their own motor-cars.'

Granting that motor-vehicles are likely to supersede both tramways and horse-vehicles, what are really the prospects of their superseding railways, as well? Should railway shareholders at once sell out and put their money, preferably, in motor-omnibus and commercial motor companies?

In regard to goods we have the fact that the quantities thereof carried by the railways of the United Kingdom in 1910 were:

Minerals	405,087,175 tons.
General merchandise	109,341,631 tons.
Total	514,428,806 tons

Motor transport could obviously not be adapted to the transport of 400,000,000 tons of minerals, and for these, at least, the railways would still be wanted. But the number of motor-vehicles necessary to deal with 109,000,000 tons of general merchandise would still be prodigious, apart from considerations of distance, time taken in transport, wear and tear of roads, and, also, of the question whether a locomotive, doing the work of many motors, would not be the cheaper unit in the conveyance of commodities carried in bulk on long or comparatively long hauls. The suburban delivery of parcels is one thing; the distribution, for example of 1000 railway waggons of broccoli from Penzance, all over Great Britain, in a single week, is another.

In the matter of passenger traffic, while people of means may prefer to make such journeys as that from London to Scotland in their own motor-car, the railway will continue to form both the cheaper and the quicker means of travel for the great bulk of the population as distinct from private car-owners, whose number must needs be comparatively small.

It is in respect to urban and suburban traffic that motor-vehicles have their best chance of competing with the railways on any extensive scale; yet even here, and notwithstanding all that they are already doing, their limitations are no less evident.

Taking only one of the many railway termini in London, the average number of suburban passengers who arrive at the Liverpool Street station of the Great Eastern Railway Company every week-day (exclusive of 12,000 from places beyond the suburban district) is 81,000 and of these about 66,000 come by trains arriving, in rapid succession, up to 10 a.m. To convey 81,000 suburban dwellers by motor-omnibus instead of by train would necessitate 2382 journeys, assuming that every seat was occupied. On the basis of the average number of persons actually travelling in a motor-bus at one time, it would probably require 4000 motor-bus journeys to bring even the Great Eastern suburban passengers to town each day if they discarded train for bus, and the same number to take them back in the evening. So long, too, as a single locomotive on the Great Eastern suffices for a suburban train accommodating between 800 and 1000 passengers, the company are not likely to pull up their rails and provide tracks in their place for a vast 'fleet' of motor-cars or motor-omnibuses.

In some instances tramways and motor-omnibuses have, undoubtedly deprived the railways of considerable traffic, and certain local stations around London have even been closed in consequence. In other instances tramways and buses have been of advantage to the railways by relieving them of an amount of suburban traffic for which it might have been difficult for them fully to provide. But any *general* supplanting of railways by motor-vehicles is as improbable in the case of passenger travel as it is in that of goods transport. Motor-vehicles are certain to become still more serious rivals of the railways than they are already, but they are not likely to render them obsolete; and, taking the country as a whole, the 'bulk of the traffic' may be expected still to go by rail, motor-vehicles notwithstanding.

Although, at the outset, some of the railway companies were disposed to regard the motor as a rather dangerous rival, the most enterprising have themselves adopted various forms of motor-vehicles, alike for establishing direct communication between country stations and outlying districts unprovided with branch lines, for enabling passengers arriving in London to pass readily from the terminus of one company to that of another, and for the collection and delivery of goods.

In regard, again, to the outlook for the future, important possibilities were foreshadowed by a letter addressed to *The Times* of August 23, 1911, by Lord Montagu of Beaulieu, concerning 'Road Transport during Strikes.' The hope of the leaders of the then recent railway strike had, of course, been to produce such a paralysis in the transport arrangements of the country that the railway companies would have been forced, owing to the resultant loss, dislocation of traffic, and, possibly, actual famine conditions, to surrender to all the demands made upon them. While the attempt failed on that occasion—thanks to the loyalty of the majority of the workers, the almost complete lack of public sympathy with the strikers, and, also, the employment of troops for the protection of the railways—there will always be the possibility of a renewal of the attempt. Pointing, therefore, to the large number of motorists in the United Kingdom, and mentioning, also, that there are, in addition, at least 10,000 commercial motor-vehicles as well, mostly running in or near the larger industrial centres, Lord Montagu wrote that, if supported by the Royal Automobile Club and the Automobile Association and Motor Union and assisted by his brother motorists in general, he would undertake in the case of a national emergency to carry out the following operations:

1 The carriage of all mails where railways are now used.
2 The supply of milk, ice and necessaries to all hospitals and nursing homes.
3 The supply of milk, fish and perishable produce to London and other large towns.
4 The supply to country villages of stores not produced in or near their area, such as sugar, tea, etc.
5 The carriage of troops or police.
6 The conveyance of passengers if on urgent business in connection with family matters or trade.

Lord Montagu added that 'the Government would, of course, have to guarantee open roads and protection for loading and unloading vehicles, and provide for the swearing-in of motorists as special constables, who would be thus engaged in saving the community from starvation and chaos.' He further thought that the compilation of a national register of motorists willing to lend their cars should be proceeded with at once.

The existence of such an organisation as this, with the inclusion, also, in the proposed registry, of horsed waggons, waggonettes and other vehicles owned by the country gentry and others, might be of incalculable service both in enabling the railway companies to stand against the coercion of a really general strike, and in saving the transport of the country from any approach to a complete dislocation, pending the time when the full railway services could be resumed.

A further example of the possible usefulness of motor-vehicles was shown by a War Office memorandum, issued on September 26, 1911, giving particulars of a provisional scheme for the subsidising of petrol motor-lorries already manufactured and owned by civilians, complying with certain specified conditions, the War Office thus acquiring the right to purchase such lorries from the owners for military service, in the case of need.

Measures of the kind here in question would, of course, be temporary expedients only, there being, as shown above, no probability that motor transport by road would ever take the place altogether of transport by rail.

Nor is aerial locomotion likely to be a more formidable rival of the railways than either inland navigation or motor transport by road. One may safely anticipate that further great advances are yet to be made in the art of flying; yet one may, also, assume there is no prospect of aerial locomotion becoming a serious competitor with the railway. It is extremely interesting to know that the journey from London to Scotland has now been made in quicker time by aeroplane than by the fastest express, and that a 1000-mile flight round England has been accomplished with perfect control of the machinery employed. Yet, even allowing for the greatest possible improvements in the construction of the aeroplane, the number of passengers who could be carried is so limited, and the fares charged to cover capital outlay must needs be so high, that there could be no idea of rivalry between the aeroplane and the railway in regard to passenger traffic.

Like considerations should apply in the case of goods traffic.

In theory the idea of an aerial express goods service looks very promising. Yet, as a business proposition, one must needs again consider: (1) the capital cost of the aeroplane; (2) the comparatively small quantity of goods that could be carried on a single journey; and (3) the high rates that would necessarily have to be paid for their transport on commercial lines. A 'record' in the aerial carriage of a 38-lb. consignment of electric lamps from Shoreham to Hove (Brighton) was established on July 4, 1911, by Mr H.C. Barber, of the Hendon Aviation Grounds; but this particular exploit was suggestive mainly of an advertisement for the lamps in question. I ventured, therefore, to put the following proposition to Mr. Barber:

'Assume that, owing to a railway strike, no goods trains could pass between London and Liverpool, and that a London merchant had a consignment of goods which it was of the utmost importance should be taken to Liverpool for despatch by a steamer on the point of sailing. Then: (1) What would be the maximum weight, and, also, the maximum bulk, of such consignment as an aeroplane could carry? (2) In what time, approximately, could the journey from Hendon to Liverpool be made? (3) What sum would the London trader have to pay for the transport?'

Mr Barber informs me that the maximum weight of such consignment as could be carried would be about ten stone (1 cwt. 1 qr.); that the maximum bulk would be about 30 cubic feet; that the journey would take about four hours; and that the charge for transport would be ten shillings per mile. The distance 'as the crow—or the aeroplane—flies' between Hendon and Liverpool being about 200 miles, the charge would come to £100. Mr Barber adds: 'There is no doubt that within the very near future it will be possible to make much smaller charges; also charges could be very much reduced if there were sufficient business to make it worth while.' This is what one would expect to hear. Yet, assuming that the aeroplane rate were reduced even by fifty per cent, it could not, even then, compete with the railway rate under normal conditions; while to convey through the air the 150 tons of general merchandise which a single locomotive attached to one of the many goods trains passing between London and Liverpool will haul would, on the basis of 1 cwt. 1 qr. per machine, require the use of 2400 aeroplanes. This calculation leaves out of account, too, the much greater weights of grain, timber and other heavy traffic in full truckloads which pass from Liverpool to various inland places, and could not, of course, be dealt with by aeroplane at all.

After surveying all these possible competitors or alternatives we are left to conclude that, as far as foresight can suggest, the railways are likely still to constitute at least the chief means of carrying on internal transport and communication in this country. [. . .]

While the outlook for the future has various elements of uncertainty, and, in regard to matters of detail, gives rise to some degree of concern, a review of the conditions under which trade, industry and communication have been developed throughout the ages leads to the conclusion that the country may, at least, regard with feelings of profound thankfulness and generous appreciation the efforts of that long succession of individual pioneers, patriots and public-spirited men to whose zeal, foresight and enterprise we are so materially indebted for the advantages we now enjoy.

Notes

1 A good example of these tendencies is offered by the Southend district, situate at the mouth of the Thames, a distance of 35 miles from London. Season tickets between London and Southend are issued by the railways at a low rate, and on the London, Tilbury and Southend line there are 6000 holders of these tickets. In the special interests of wives and daughters cheap tickets to London by an express train are issued on Wednesdays to allow of shopping in town, visits to the theatre, etc., and by this train there is an average of from 600 to 700 passengers, consisting almost exclusively of ladies.
2 Exclusive of season-ticket holders.

4.2 H. A. Osgood, *Transportation*, 1937

The importance of transportation

Whether or not we agree with Kipling's assertion that 'transportation is civilization', it is plain that most of our present civilization is dependent

Source: H. A. Osgood, 'Transportation' in National Resources Committee, *Technological Trends and National Policy including the Social Implication of New Inventions*, US Government Printing Office, 1937, chapter 3.

on transportation for its existence and that the transportation industry itself is one of the most important factors in the economic and social life of the United States.

The Federal Coordinator of Transportation recently found over $27,000,000,000 invested in rail, pipe-line, and waterway transportation in this country. To this investment at least $2,000,000,000 may be added for motortrucks, many billions more for privately owned automobiles, and correspondingly great sums for improvements of inland waterways, rivers and harbors, and our 3,000,000 miles of highways. Air transportation is an industry in which millions in new capital are being added annually.

Similarly impressive statistics as to employment, purchasing power, taxes paid, etc., are readily available and, in a general way, familiar to all who have given even the most casual attention to the subject.

Probably any extended discussion of the part transportation has played in our history would verge on 'a blinding glimpse of the obvious.' Economically the United States is the world's greatest experiment in free trade within its borders, spread within our continental limits over about 3,000,000 square miles of land of the widest diversity. That our 48 States could ever have been developed or united, or that the present Nation could be held together without a great and efficient transportation system, is highly improbable.

Our present transportation service

In attempting to ascertain the present scope of freight and passenger traffic in this country, the investigator ventures into a strange and terrible jungle of statistics. Not only are the figures of a size usually associated with astronomical calculations, but a regrettably large amount of the basic data would have delighted Mark Twain's Connecticut Yankee, who discovered that in King Arthur's day one merely had to state one's facts and was not required to prove them.

Railroad statistics are compiled accurately and in expensively elaborate form; waterway figures are accurate and detailed; but do not include ton-mile data for coastwise or intercoastal traffic, and do include a large amount of short-haul intraport traffic at places like New York and San Francisco (the equivalent of a railroad switching service); pipe-line figures have to be adjusted arbitrarily for intrastate traffic; while motor-transportation statistics are scarcely to be dignified by such a label. Ton-mile figures for motor-truck traffic, for example, have ordinarily been derived from an assumption that 50 percent of truck-miles are purely local, an estimate that empty mileage of trucks amounts to 33 percent, a conjecture that trucks on the average load to 80 percent of their capacity, and a guess that the average truck makes a certain arbitrary mileage per year. Wherein such a mathematical guess is superior to the ordinary or inspirational guess is problematical.

Within fairly broad limits, and following mainly the study of the Federal Coordinator's Freight Traffic Report, freight service in the United States appears to be divided as follows:

Freight traffic in the United States, 1932

	Billion ton-miles	Percent
Railway	236	69
Pipe line	32	9
Water	37	11
Highway	35	11
	340	100

The passenger traffic situation, based entirely on the Federal Coordinator's reports, is as follows:

Passenger traffic in the United States, 1933

	Billion passenger-miles	Percent
Common carriers:		
Railway	16.3	4.3
Bus	3.4	.9
Airway	.2	.1
Total	19.9	5.3
Private automobiles:		
Intercity	184.9	48.9
City	173.0	45.8
Total	357.9	94.7
Grand total	377.8	100.0

Probably it would be even better to say that the railroads perform about two-thirds of the freight service, and that the balance is divided fairly evenly between highways, pipe lines, and waterways.

As a result of the inaccuracies and lack of similarity in data, such wide differences of opinion as usually arise from uncertainty as to facts have grown up and even been reflected in graphic charts and expert reports prepared for the enlightenment of students of the transportation problem and for the public in general.

Grand total

Here again it is plain that with 24,200,000 private automobiles registered in the United States, and completely free from any statistical obligations whatever, the necessary facts as to total automobile mileage per year, the number of passengers carried in addition to the drivers, and the division of travel between intercity traffic on the one hand and going to the corner grocery or the neighborhood movie on the other are totally lacking. Probably here it should be enough to say that the overwhelming bulk of our passenger transportation is performed by private automobiles. The division of the remaining 5 or 10 percent which is handled by common carriers seems accurate.

Twenty years ago the Supreme Court condemned some exhibits before it as an example of 'setting down figures with delusive exactness.' No better characterization could be found for many transportation statistics.

The general future of transportation

In considering the future of transportation over the next 20 or 25 years it will be advantageous to deal first with freight and passenger traffic as such and without regard to the individual agencies performing these services.

Freight

It appears unlikely that freight traffic measured by ton-miles will increase materially beyond the standards which prevailed prior to 1930. In the first place, we use a great deal more freight service per inhabitant than other countries do; 1933 figures (except for France and Switzerland, where the averages cover the year 1934) are given by the Bureau of Railway Economics as follows:

Country:	*Ton-miles (railway)* *per inhabitant*
Great Britain	430.25
Germany	526.21
France	590.36
Sweden	368.72
Italy	141.18
Switzerland	282.69
India	77.44
Japan	133.12
Canada	2,247.37
United States	2,137.96

Granting that the proportion of motor truck and waterway service in these countries may differ somewhat from our own, and making whatever allowance seems reasonable for differences in area and density of population, the United States still seems to use a disproportionate amount of freight service as compared with older and more settled and stabilized countries.

The trend of our population is toward stabilization. The Census Bureau thinks that 11 of our 48 States have fewer inhabitants than in 1930. We have various persuasive historical studies of our vanished frontier and its profound effect; we have reached the stage where governmental efforts are being made to retire land from cultivation and settlement. The United States still has room for many million additional inhabitants, but beyond the horizons which are new to them are dude ranches and hard roads.

The second factor limiting the growth of freight transportation is not advanced science and revolutionary inventions, but prosaic common sense stimulated by competition. Everywhere the industrial world is trying to eliminate useless transportation, and the cumulative effect is already noticeable. A good instance in the not distant past has been the production of steel, using hot metal direct from the blast furnace and eliminating the transportation of pig iron and of the fuel formerly used to melt it. Other examples are the shift of the textile industry from New England to the Southern States, and of the shoe industry from the East to the Middle West. Location of industries where freight costs can be saved is a conspicuous factor. Manufacturing plants are being built out of savings in freight rates. In lighter manufacturing particularly an undoubted tendency toward relocation and dispersion of industry is evident, although this tendency does not arise from transportation considerations alone.

Probably even more important than such factors over a long term of years is the elimination of waste, not merely waste freight transportation arising out of bad location of industry and from circuitous hauls but waste of materials. Conspicuous examples are the waste of coal in domestic heating and small power plants, evidenced by our smoke nuisance, waste of gasoline through bad carburetor adjustments, etc., and waste of oil burnt under boilers at a fraction of the efficiency at which it might be used in internal-combustion engines. It is not necessary to review the literature on this subject—for over a generation we have all been schooled to believe that the United States is the most wasteful Nation in the history of the world, and the technical aspects of the situation were given wide publicity by former President Hoover. Nor need we argue whether so great a Nation as ours could have been developed so rapidly without a large if not a disproportionate amount of waste.

Any study of freight transportation of the future, however, must recognize that if, say 20 or 25 percent of our materials are being wasted, we have a corresponding amount of freight transportation awaiting elimination.

This naturally leads to the third factor limiting freight transportation—advanced technology, including, of course, the use of materials at present wasted, but particularly affecting traffic through the production of better materials and the manufacture of better designs. Developments along these lines in the fields of agriculture, mining, the chemical, and metallurgical industries in the production, transmission, distribution, and uses of power, etc., are treated in the various other special and general papers comprising this entire study of technological trends and their social implications.

Freight traffic is peculiarly a product of factors beyond its control, and in a survey of such factors must be read the future of the transportation of freight—both as to its volume and its character. We have better and longer-lived materials. We ride on tires advertised to run 20,000 miles, and coming much nearer to meeting this standard than the guaranteed 3,000-mile tires of 20 years ago came to the claims of their manufacturers. The American Iron and Steel Institute calculates that the 34,000,000 tons of steel produced in 1935 may be expected to last an average of 32 years, or approximately twice as long as steel did 40 or 50 years ago. The development of alloy steels, minimizing rusting and making the steel itself stronger and more durable, improved manufacturing eliminating impurities, improved processes for coating steel products with tin and zinc to resist corrosion, and the refinements in manufacturing processes and rigid tests assuring higher quality and insuring fewer replacements once the steel is in use, are pointed out by the Iron Age as factors in lengthening the life of steel.

As in material, so in designs—particularly designs permitting the use of higher speed, lighter machinery. This general field is far too broad to serve as part of a special report on transportation—plainly, however, any increases in freight traffic do not lie in this direction.

A fourth factor, not merely limiting freight transportation but making constant, direct inroads into the volume of freight service required, is the competition of electric transmission lines and natural-gas pipe lines.

Coal has long been the backbone of railroad tonnage; it is a large element in water transportation, and in many areas it is handled in great quantities by motortrucks. In 1928, a fairly typical year, for instance, about a third of all the tonnage moved by the railroads was coal, which also produced about a fifth of all the freight revenues. Cutting into this transportation of fuel is the electric-power industry, employing a fourth as many persons as the railroads operating over 200,000 miles of transmission lines, transmitting around 100,000,000,000 kilowatt-hours of electric energy annually and striving to minimize the haulage of its own fuel. Fuel for this industry does, of course, move in tremendous tonnages, and these tonnages may increase with increased use of steam-generated electricity.

It is true also that generating plants cannot ordinarily be located at the mines on account of the lack of an adequate supply of cooling water. Beyond certain distances electric energy cannot be transmitted cheaper than coal can be hauled; electricity cannot be stored in wholesale quantities, and coal can.

Subject to these limitations, however, electric-power companies naturally try through the location of their plants to reduce haulage of coal. Plainly, improvements in electric transmission may decrease transportation of coal. The amount of coal required per unit of electric output is being steadily reduced.

Moreover, as about one-third of the electric energy generated by public utilities is now developed by water projects as Boulder, Grand Coulee, and Norris, the transportation of coal will be further affected. Some of these projects serve portions of the country where cheap coal is not available and will involve little direct displacement of coal, the fuel replaced, if any, being oil or gas. Others, however, will tend to limit the transportation of coal. It seems improbable that the national requirements for bituminous coal will exceed the 1929 level for some years to come.

Less known to the general public is the growth of natural-gas trunk pipe lines. In 1931 there were about 70,000 miles of these and probably $3,000,000,000 is invested in this industry, which is said to have more than 5,000,000 domestic customers. Forty-four percent of the production of natural gas is used in drilling oil or gas wells, or for the manufacture of carbon black. The balance represents direct substitution of gas for other fuels, chiefly coal and oil—equivalent to over 40,000,000 tons of coal a year, and the industry is progressing at a rate undreamed of by the average person.

While freight traffic, measured by ton-miles, will probably increase but slowly above the normal levels of the past, lighter and bulkier freight will in some measure serve as an offset from a revenue and even from a carload or truckload standpoint. Such items as electric refrigerators and radios have reached a surprising volume in recent years. A large traffic in fruits and vegetables, from such distant territories as the Rio Grande and Imperial Valleys, has grown up, and the near future will probably see increasing movements of air-conditioning and insulating equipment and materials, portable houses, trailer bodies, and the like. These increases in volume will partly compensate for reductions in weight.

While no great, new heavy industry has appeared on the horizon, many new lighter industrial and manufacturing activities are in plain view. This trend will call primarily for flexibility and speed in transportation.

On the whole, however, the gradual increase of population appears as the principal favoring factor in the freight situation. Against this may be set off some of the most vital influences in our economic life.

Passenger

The probable future of passenger transportation presents a striking contrast to the freight situation. The urge to travel is undoubtedly a deep-seated human characteristic. Not only has it been evident from the earliest times down to date but a yearning for 'fresh fields and pastures new' is apparent through the widest ranges of social classes. No more striking example of this general thesis can be found than in the United States. Here the average travel per inhabitant was about 500 miles per annum in 1920, and over 2,000 miles per annum in 1929.

In the words of the Federal Coordinator (from whose Passenger Traffic Report these figures are taken), 'within less than a decade, American travel desires and habit were quadrupled, and at the end of 4 years of depression, were still more than three times as great as they were prior to the automotive era.'

In a paper read before an international congress for traffic problems at Vienna on June 16, 1936, the director of the Austrian Governmental agency for the development of tourist traffic predicted that the next few decades will bring a gigantic rise in intercity and international travel, and that this mass travel—at low fares—of the near future will have to be carried primarily by the railroads as the only agency capable of handling it. These statements, based upon intense study, come from a neutral source, interested only in traffic problems as such.

All increases in leisure or in speed of travel and all decreases in transportation costs promote passenger travel, and the more people travel the stronger is their desire to do so.

Better forms of communication will take the place of some passenger traffic. The long-distance telephone has saved many business trips. If to this television is some day added, and if the costs of all such services are cheapened, communication may further supersede transportation. Such changes, however, are probably in degree rather than in character. The savage with his drums and smoke signals, the Romans with their system of flaring beacons, the mails, the telegraph, the telephone, and the radio, all have served as substitutes for travel.

Doubtless, too, a large portion of our travel is recreational or partly so, and this is not influenced by communications.

In the face of the quadrupling of our own passenger traffic in recent years, it seems improbable that communications will seriously limit travel. Ability to make more speedy and numerous contacts through improved communications may even be an item making for increased passenger transportation.

Unlike freight, where all the railroad or other common-carrier advertising, education, and solicitation in the world will not make tonnage move beyond the economic needs of the time, the passenger-transportation market is even today capable of large expansion.

Highway transportation

Probably the road passing the front door touches the lives of the people closer than any other transportation facility. Certainly the average man has both a more intensive and a wider knowledge of our highways and the vehicles using them than of any other phase of transportation. On December 31, 1935, the miles of road in the United States were as follows:

Existing types	Grand total	Total State	State primary system	Secondary roads under State control	Local roads
Nonsurfaced	2,052,263	[1]148,419	52,060	96,359	1,903,844
Surfaced	981,737	357,051	279,807	77,244	624,686
Low type...........	818,835	242,708	168,282	74,426	576,127
High type	162,902	114,343	111,525	2,818	48,559
Total..............	3,034,000	505,470	331,867	173,603	2,528,530

[1]Includes 63,628 miles fully graded and drained, and ready for surfacing.

The data given above as applied to State primary road systems and to secondary roads under State control are accurate enough, but the local road figures are not supported by equally reliable statistical information.

Primary road system

We have a well improved system of through routes traversing States, regions, and the country as a whole, connecting our principal cities and generally adequate for the traffic. It should be noted, however, that almost none of our transportation maps, whatever the agency, show density of traffic. The Lincoln Highway averaging fewer than 100 vehicles a day in Western Utah, looks just as big on the map as the six or eight-lane superhighways in the vicinity of our greatest cities.

The Bureau of Public Roads of the United States Department of Agriculture reports that main and through highways are now being designed to allow safe travel at speeds of 60 miles an hour or more. With such speeds in mind, our main roads will have smooth and skid-resistant surfaces, careful alinement, effectually moderated grades and curves, and the opposing lanes of travel should preferably be separated.

Horizontal and vertical curvature will have to be reduced to permit clear vision of not less than 800 feet on curves and at the apex of grades. Where shorter sight distances cannot be avoided, markers should be set up showing the safe speeds at which the curves can be negotiated.

In the case of three-lane highways on which the middle lane is used intermittently for passing, sight distances up to 1,200 feet are desirable.

Intersections with heavy traffic roads will require grade separation structures, less important crossings may be protected by islands in the main road so that vehicles can find protection between these islands, and accomplish the crossing if necessary in two movements through breaks in the opposing traffic stream on the main road.

Access to high speed roads will, of course, have to be regulated. Abutting property owners cannot be allowed to enter the right-of-way wherever they choose, nor be permitted to utilize their frontage indiscriminately for parking purposes or the sale of hamburgers and red china dogs. Special stopping places with suitably widened and protected shoulders should be placed along the road for the accommodation of busses and rural mail delivery cars. Safety and road utility, maintaining an unobstructed full width of pavement for through traffic, are the paramount considerations.

As traffic gravitates toward the most highly improved roads, the superior design and control of the future high speed main highway systems will lighten demands on secondary roads. Much of the cost of expensive high type construction of the main highway may thus be saved in the less costly nature of the secondary road systems. [. . .]

The vehicles
In 1935 there was one registered motor-vehicle for every 4.79 persons in the United States as compared with one for every 5.30 [. . .] in 1933. At the end of 1936, 24,200,000 passenger cars and 3,800,000 trucks were registered, both figures establishing new records and the much-discussed saturation point still not imminent.

A complete analysis of 1932 ownership was made by the Bureau of Public Roads, showing the highest degree of automobile ownership in California, where there was one car to every 2.77 people, and a generally decreasing ratio of ownership from West to East—for instance:

Colorado	3.60
Kansas	3.70
Ohio	4.05
Pennsylvania	5.72

In Arkansas one car was registered to every 13.27 people and in Mississippi one to every 12.85, as compared with one car to every 3.46 people in the State of Washington and 3.64 in Oregon. The average for the United States in that year was one car for every 5.16 people.

In the face of such figures it would be hazardous to reach any specific conclusions as to where automobile registrations will stabilize in reference to population, although the tendency of the registration curve will probably finally be to flatten out and approach a curve parallel to the curve of population.

While leaders in the motor industry are by no means fixed in their views and decidedly not unanimous, the general trend appears as follows:

Passenger cars

Size—About the same as at present which, after all, is based on the size of human beings. The small English and Continental cars are the products of taxes on large motors, the high prices of fuel and lubricants, the distances, and the character of roads.

Weight—Undoubtedly less, through the continued development of lighter metallic alloys, the possible cheapening of aluminum, etc.

Speeds—Not much beyond the performance of current 1937 models. From a standpoint of safety, a good deal of resistance to increased highway speeds has grown up, even where fast travel may be feasible. If, however, we have roads and cars permitting comfortable, economical, and safe operation at speeds of 60 miles an hour or more, we shall probably make use of such transportation facilities. Indeed, public desire to drive at high rates of speeds over modern highways, and the unwillingness of good drivers to be at the mercy of utterly unfit operators of automobiles are bound to force greater supervision of the drivers themselves. The right of 'the maim, the halt, and the blind', not to mention the financially irresponsible, the drunken, the reckless, and the miscellaneously unfit to operate motor vehicles, is not guaranteed by the Constitution or the Declaration of Independence.

A surprisingly wide public sentiment, however, seems to regard the ownership and operation of a motorcar as God-given rights quite as inalienable as 'life, liberty, and the pursuit of happiness.' [. . .]

Busses

The part played by busses in passenger transportation has been shown to be relatively small; moreover, a large number of intercity bus lines are owned by railroad companies. These railroad-owned busses frequently operate on roads paralleling the carrier's own rails, or in a nearby territory, and sometimes far beyond the carrier's lines—the Burlington Railroad, for example, operates a bus service on the west coast.

The recent formation of National Trailways for the coordination of transcontinential bus schedules, for improvement of stations, and for support of an advertising campaign, involves a number of railroads which are concerned in this enterprise through their subsidiary bus lines.

In a general way, busses provide more frequent and flexible service in light traffic territories than railroads are able to do. Certainly without plenty of passengers no road can maintain track and operate trains with an engineer, fireman, conductor, and brakeman in successful competition with a bus operated by one driver and running on a public highway.

Where busses parallel railroad operations—as between Chicago and Detroit, for a conspicuous example—travel seeks the bus largely on account of cheaper rates.

In his passenger traffic report the Coordinator said:

> To meet present-day travel requirements, carrier service should be safe, easy to procure and use, attractive, characterized by sincere and warm-hearted hospitality, well rounded and complete in every detail * * *, convenient to use and unhampered by annoying routines, as comfortable as the traveller's own home, and, finally, spotlessly clean and sanitary.

How far this is a description of the average journey by bus can best be determined by the individual reader's experiences.

Granted equal rates, the busses possess many advantages for local service but, in view of the high standards suggested by the Coordinator, do they have any advantage over the railroads on long-haul services?

Evidently this is an auxiliary and supplemental service which is working out to the decided benefit of the public and to no detriment of other common carrier forms of transportation; indeed, bus lines take the place of all branch lines and light traffic lines in bringing to railroad rails a certain amount of passenger traffic which otherwise would move by private automobile for long distances.

Motor trucks

Over $3\frac{1}{2}$ million motor trucks are registered in the United States. Their capacities may best be judged by the production figures for the past 7 years.

The table clearly shows that light trucks predominate, although it does not indicate to what extent they may be overloaded. Speed and particularly facility of operation are high factors in commercial highway transportation.

Percent truck production by capacities[1]

	1929	1930	1931	1932	1933	1934	1935
¾-ton or less	17.1	24.0	25.2	32.3	27.6	28.6	37.4
1-ton and less than 1½ ton	9.5	5.2	1.1	.6	.2	.4	.5
1½-ton and less than 2-ton	63.4	61.7	66.6	58.8	63.7	62.9	56.0
2-ton and less than 2½-ton	3.4	2.7	2.0	3.1	4.4	4.3	3.3
2½-ton and less than 3½-ton	4.1	3.8	2.7	2.4	2.2	1.9	1.4
3½-ton and less than 5-ton	1.0	1.0	1.0	1.1	.8	.8	.4
5-ton	.3	.2	.2	.6	.2	.2	.4
Over 5-ton and special types	1.2	1.4	1.2	1.1	.9	.9	.6
Total	100.0	100.0	100.0	100.0	100.0	100.0	100.0

[1]Compiled by Automobile Manufacturers Association.

Based on the foregoing table, the average truck has a capacity of about $1\frac{3}{4}$ tons, and the 3,511,000 trucks registered in 1935 had an aggregate capacity of about 6 million tons.

In estimating the influence of trucks on railroad freight service it should be noted that the freight-carrying car capacity of the railroads is over 105 million tons, or about 17 or 18 times the aggregate capacity of all trucks in the United States. It should also be borne in mind that a large portion of the motor trucks is used for city delivery service and on farms—more than three-quarters of all trucks being Chevrolets, Dodges, or Fords—and none of this service being competitive with the railroads.

At the same time the influence of the truck is not to be minimized. Forty-eight percent of livestock receipts in 17 leading markets, 56 percent of egg receipts at Chicago, 98 percent of milk moving to 19 principal cities, large quantities of fruits and vegetables, furniture, rubber tires, short-haul coal, grain and grain products, bakery goods, and, of course, automobiles themselves, and, particularly, general merchandise, are all large items of truck traffic.

Furthermore, truck operators are, to a large extent, selective as to the commodities they handle—they do their best to skim the cream. In the past few years, where the production of heavy durable goods has fallen off so markedly, the inroads made by trucks have been more serious than they would have been under normal conditions.

In a general way trucks may be said to go after revenue rather than tonnage; to seek finished and manufactured materials rather than raw materials; and to handle consumer goods rather than capital goods.

A questionnaire of the Federal Coordinator to determine why shippers selected trucks in preference to railroad service gave the following results. (In most cases more than one reason for using trucks was given, hence the percentages below do not total 100.)

	Percent
Store door delivery	65
Faster service	65
Cheaper total cost	53
Store door pickup	51
More flexible service	43
Cheaper packing	21
Late acceptance of shipments	21
Simpler rate classification	16
Less loss and damage	11
Personal interest or friendship	3

A department store in New York may ascertain from the Weather Bureau Friday afternoon that Saturday will probably be a rainy day. In the Saturday morning papers the store will advertise a special sale of overshoes and rubber coats, telephoning a manufacturer, probably 150 miles away, Friday afternoon to furnish the sizes and styles desired. The shipment is loaded out that night in trucks and delivered at the store's door early enough the next morning to permit display before the advertised hour of the sale. Such a direct, speedy, personalized service is more than the average railroad can offer.

Ordinarily a railroad has to run trains to suit the convenience of a community and not the individual desires of some particular shipper. [. . .]

Railway transportation

Motive power

In 1936 the class I railroads owned 45,000 locomotives, over 98 percent of them being steam engines. Over 60 percent of the railroad locomotives are more than 20 years old.

Looking to the future three types of railway motive power will probably be used—electric, Diesel, and steam.

The recent electrification of the Pennsylvania–New York–Washington lines, including particularly the great terminals in New York and Philadelphia, less recently the electrification of the Illinois Central suburban zone out of Chicago and the N.Y.C. electrification of the Cleveland terminals, together with the older electric lines of the New

York Central and the New Haven out of New York, have given millions of people personal experience with this form of transportation. Invariably they are impressed with the electrified road's rapid acceleration, smooth operation at high speeds, cleanliness, and ability to handle anything from the smallest switch locomotive or single unit passenger car to the longest, heaviest, and fastest freight and passenger trains.

Railroad electrification, however, is the result of necessity rather than desire. This necessity probably arises first in the operation of congested terminals, particularly where on account of tunnels (New York is the most conspicuous example) no form of combustion locomotive can be used. Popular demand for smoke elimination is a factor, but an electrified terminal's ability to handle traffic beyond the economical and often beyond the possible capabilities of steam equipment is the principal reason.

Main-line operation with electric locomotives plainly has many advantages once terminals are electrified. Here again the justification of electrification rests far more on density of traffic than on any other factor.

Electrification involves a heavy additional investment, probably creating little additional traffic. Only with a large volume of business can enough operating expenses be saved to justify the increased capital charges. Broadly speaking, a railroad line will have to double its capitalization in order to electrify. It is true that the electrified road will cut its coal bill in two and effect other less striking operating economies, but obviously the new capital costs must be spread over a tremendous volume of freight and passenger business.

[. . .]

Rail versus highway
For reasons previously outlined, freight traffic as a whole is likely to return slowly to levels existing prior to 1930, and beyond these levels will grow even more slowly.

The graphic chart, comparing railroad freight service with various indices of general business activity, shows a remarkably close relationship over the period 1919–30 and a disappointingly slow increase in revenue ton-miles since 1930–31.

It is true that the railroads in 1934–35 originated only about 50 percent as great a tonnage of durable goods as they did in the years 1923–29, inclusive. Even assuming, however, that much of this loss in the heavy commodities will eventually be recovered, the trend shown by the chart still leaves much to be desired from a railroad standpoint.

An underlying factor of great importance arising out of stabilization of population and the settling up of large areas, is the growing commercial demand for rates and service based on something between the present less-than-carload shipments and the full carload. This demand was manifest in southern New England 25 years ago, and evidently increases as population tends to become static. The commercial requirements of many shippers are undoubtedly for smaller quantities than present railroad carload minimum weights.

Maintenance of minimum weights proportionate to the ever-increasing size of box cars and failure to establish rates based on shipments in quantities a fourth or a fifth of the full tonnage a car can handle have contributed in no small measure to the building up of motortruck traffic.

Added to this, on short-haul high-speed merchandise traffic the motortruck is in many ways definitely superior to the railroad. Ability to pick up and deliver freight, more liberal packing requirements, and a service both more speedy and far more individualized to the particular shipper's needs, are obvious to anyone dealing both with motortrucks and railroads.

The solemn progress of a few hundred pounds of freight loaded in a 40-foot box car from one small New England town to another is vaguely reminiscent of the interest of a circus elephant in a solitary peanut.

On the other hand, with heavier loads and longer hauls—in doing what would be termed a wholesale rather than a retail business—the advantage is almost entirely with the railroad. As the Federal Coordinator recently stated:

> I now see little future for long-haul motortruck haulage of most commodities, although I expect to see the shorter-haul operations expand continually.

Plainly, however, a definition of long haul is what is needed.

Mention has been made of a demand for equipment, rates, and service somewhere between the less-than-carload shipment and the full-carload shipment. It is in this zone, as well as in the zone between long-haul and short-haul traffic, that railroads and trucks are in the most intense competition.

Based partly on the old principle 'If you can't lick 'em, jine 'em', and partly on the even older idea that distant pastures are always green, attempts are being made to coordinate rail and truck service. The idea, of course, is for the trucks to handle terminal operations, but for the line haul between one city and another to be handled by the railroads, keeping trucks off our intercity highways and turning over to the railroads the strictly haulage or 'line' service.

Among present developments are freight containers which may be loaded by shippers, hauled to railroad freight stations, placed on flat cars and further shifted at transfer stations if necessary to provide for solid carloads of containers to a particular destination, where, in turn, the containers are unloaded and trucked to the ultimate receivers of the freight.

Or coordination may take the form of gathering freight by trucks and transporting the truck bodies themselves by rail from one city to another. If the truck haul be lengthened beyond the terminal zone, and such arrangements are ultimately found profitable and approved by public authority and public opinion, a railroad may thus invade the territory of its competitors and intensify a competition which has in many instances already been carried to unsound lengths.

Whether the trucking facilities in such cases are owned by the railroads, or whether through rates are divided between the railroads and independent trucking concerns, is probably immaterial from the public standpoint.

It should be noted, however, that no railroad, much less 8 or 10 different railroads, can possibly make arrangements of this nature with all the truck operators in New York, for instance. Whether a road owns and operates its trucks or whether it makes contracts with any reasonable number of truck operators, it seems doubtful that such a plan would fit the individual shipper (who, of course, in many cases owns his own trucks) so precisely and satisfactorily as do present arrangements. It is also a question at just what distances line-haul traffic can be more economically handled by railroads than by trucks which are already loaded for particular destinations.

Containers, of course, presuppose a rather limited number of trucking agencies, railroad-owned or otherwise, but present the same essential problem, that of cost. Incidentally, any development of container traffic will probably involve smaller units in addition to sizes which are now used.

Certainly this is a field in which prophecy should not be too positive. Probably it is enough to point out that from the standpoint of the shipper and also from that of the general public the prospects of increased flexibility, increased speed, more competition, and lower costs, are all pleasing.

Again quoting the Federal Coordinator's report:

> Whether regulated or not, the prices charged by highway or waterway operators, as well as by railways, inevitably will be limited to what it costs the shipper to provide his own service as a ceiling, and what it costs the carrier to furnish it as a basement * * *. The ability of the shipper to provide his own transportation by highway limits the operation of the value of the service principle to the cost of such owner transportation, as the maximum charge which may be imposed.

Where the Government regulation may ultimately step in to the benefit of all concerned is in preventing the trucks and railroads from cutting each others' throats. Government regulation in the past has probably saved a good many roads from attempted suicide, and this particular sort of history may quite easily repeat itself among the trucks. [. . .]

Air transportation
Transport operations
In the year 1926, 5,800 passengers were carried in the course of regularly scheduled air transport operations in the United States: our airlines are now handling over 100,000 passengers per month. How far the 5,800 hardy pioneers traveled is unknown; the average journey today is in excess of 400 miles. In 1937 the passenger miles flown on transport planes will probably exceed half a billion.

In 1926, 800,000 pounds of mail were carried, and in 1935 over 13,000,000 pounds.

Figures like these, marking the rapid development of civil aeronautics in the United States, can be compiled to wearisome statistical lengths. The amazing growth of aviation, the novel and spectacular nature of the industry, and the realization of man's age-old dream of flight have combined to put flying first in the public mind whenever transportation is discussed.

Probably popular imagination as to this form of transportation is somewhat overstimulated. Passenger-miles flown in the Continental United States in 1933 were 173 million, as against 16 billion passenger-miles handled by the railways and 185 billion by private automobile in intercity travel. While it is true that the transport plane passenger-miles almost doubled between 1933 and 1935, and will quite probably double again before 1940, we are still talking about a transportation agency handling only a fraction of 1 percent of the business.

We are also discussing a form of transportation that, while in many ways only on the threshold of its development, already faces definite fundamental limitations.

Short-haul traffic as between New York and Philadelphia is, of course, out of the question—from the standpoint of time alone the plane is inferior to the train. On longer hauls, as between New York and Washington, air travel may save an hour out of $3\frac{1}{2}$ hours, but for most people this is not enough to justify the safety factor and particularly the higher fare. Where airports can be relocated closer to large cities, air transport will profit correspondingly.

In addition to regularly scheduled transport operations, which in 1935 involved about 350 planes in domestic service and 100 in foreign, we had about 8,500 licensed and unlicensed other airplanes in operation carrying a million passengers. A great many of these people merely made 10-minute sightseeing flights at airports on pleasant week-ends, but the figures also include strictly transportation flying.

Quite a number of companies and individuals own airplanes, and some flying is done in specially chartered planes, athough the average rate for the latter sort of service is about 20 cents a mile. Charter operators carry passengers on flights which are not made on scheduled airlines—for example, from a city which does not have an airline to one which does, or at some particular hour.

Unless one has closely followed the amazingly rapid developments of aviation and generally unless he has actually flown, it is highly improbable that the average man realizes the present stage of civil aeronautics in the United States.

Air liners going into service in June 1936 are 21 passenger bimotored low-wing monoplanes with top speed of 215 miles an hour.

Air lines are now asking for a plan capable of flying 900 miles nonstop with 40 passengers, or 2,000 miles nonstop with 20 passengers and berths. Such a plane is to use four engines, each developing 1,000 horsepower at sea level and 900 horsepower at 9,000 feet altitude. A top speed of 230 miles an hour is asked, cruising speed of 210 miles an hour at 75 percent engine rating, and 193 miles an hour at 60 percent horsepower.

That present passenger planes rival any form of transportation in comfort and even luxury, and far surpass other agencies in speed, is generally accepted by the traveling public.

Unlike railroads, which have to own and maintain their lines and terminals, the transport airplane to a large degree depends upon Federal aids to air navigation and, of course, on the aid of cities for terminals. At the end of 1935 our principal airports were owned as follows:

Commercial and private	552
Municipal	739
Intermediate—Department of Commerce:	
Lighted	282
Unlighted	9
	1,582

"WHAT WILL HE GROW TO?"

1. A *Punch* cartoon of 1881 proclaiming the birth of a new industrial prime mover.

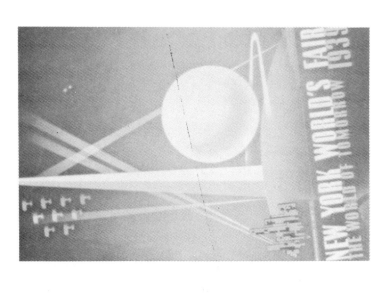

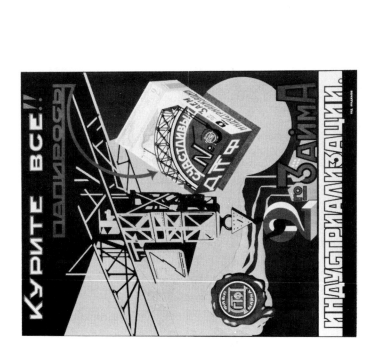

2. Two posters using modern design to promote new technology, albeit within two contrasting economic systems. On the right a poster for the New York World's Fair of 1939. International expositions remained a shop window for new technology throughout the period. Apart from the modernist architecture depicted in the poster, this one was notable for General Motors' 'Futurama' exhibit, foretelling a technological utopia. On the left, a Soviet cigarette poster exhorting everyone to smoke a state brand (*Schastlivy*, meaning Happy) to help finance industrialization. It was common for cigarettes and other consumer goods to have industrial names and motifs; this poster dates from 1928, the start of the First Five-Year Plan.

Have you discovered Electricity?

The successful efforts of an ancient philosopher to produce ELECTRICITY by rubbing amber remained one of the curiosities of human knowledge for many centuries.

The last few years of scientific progress have sufficed to bring ELECTRICITY to your aid in daily life and work.

To-day a vast Industry exists to provide YOU with Electric Power and the appliances for its use in YOUR Home.

Have YOU discovered that the new Electric Service offers you ease, efficiency and health in YOUR life.

Consult your Local Suppliers of ELECTRICITY and Appliances.

Issued by
THE BRITISH ELECTRICAL
DEVELOPMENT ASSOCIATION,
15, Savoy Street, Strand, London,
W.C.2.

3. Have You Discovered Electricity? *Electrical Age for Women,* 1 (1926–1930), p.41. The British Electrical Development Association, founded in 1919 to educate the public about the uses of electricity, devoted considerable effort to promoting domestic electricity, stressing in this case the progressive role of science-based industry in transforming the life of the houswife/consumer.

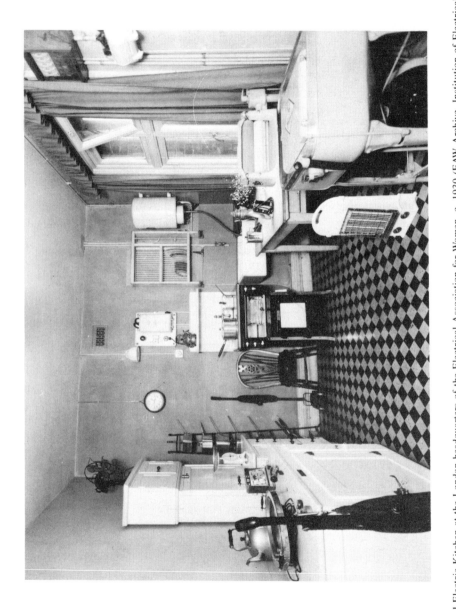

4. The All-Electric Kitchen at the London headquarters of the Electrical Association for Women, c. 1930 (EAW Archive, Institution of Electrical Engineers). Here women could learn how their kitchens might be adapted for electricity, using an impressive array of appliances. Note the surface wiring in conduits.

5. 1930s aerial photograph of East River, New York City, depicting the impact of new materials on the metropolitan built environment. The farther of the two bridges, the Brooklyn Bridge, completed in 1883, was one of the first to use Bessemer steel in its superstructure; and it was bulk steel which helped make the skyscraper possible.

6. London Underground poster, 1908, indicating links between transport innovations and new residential patterns. Golders Green, six miles north-west of Central London, was little more than a hamlet in 1907, when it became the first place outside the London County Council boundary to be reached by the electrified Underground railway. By 1931 its population had risen to nearly 18,000.

7. Women Consumers. *Left,* Lux, Cleansing Solution, Ad. 1902. This is an early example of the modern female mobile shopper, clearly travelling quite a way by car to purchase a particular branded soap product to bring higher standards of cleanliness to her family. *Right,* 'As dependable as an Austin', 1930. In America, women were quick to adopt the new form of transport, and they were targeted as car users in Britain as well.

8. Two examples of new communications media in everyday life. *Above*, an Ecko radio, model AD65. The bakelite cabinet, designed by Wells Coates in 1934, marks an attempt to use new materials in a way unconstrained by traditional furniture design. *Below*, a new kind of urban congestion; overhead telephone wires at Pratt, Kansas in 1909.

9. An instance of the power of modern communications to shape perceptions of the world. When U.S. President Theodore Roosevelt forgot his promise to take a Selig camera on his African hunt in 1909, Colonel William A. Selig 'nature-faked' a screen version at the Selig Polyscope Company's studios in Chicago, using a vaudeville actor to impersonate Roosevelt. He did, however, use a real lion.

10. The Right Flour, 1906. In the quest for ever whiter flour to satisfy public taste, various dietary elements now recognized as important were removed, even when home-baking was still the norm.

11.   The Mechanization of Food Production. *Left*, an early tractor in use in Britain. It is clearly an early design, showing the machine's origins in the automobile. This one is substituting for horse draughting power, but in Britain until the 1940s, tractors were more commonly used to substitute for steam as horses continued to be economic. *Right*, women sorting beans for H.J. Heinz & Co., Ltd., Harlesden, c. 1929, the mass-production of processed foods for mass consumption.

12. Two aerial views capturing the effects on housing of industrial and administrative changes. *Above*, typical nineteenth-century housing for industrial workers at Preston in the north of England. *Below*, a surburban housing estate at South Croydon on the outskirts of London (picture taken in 1948).

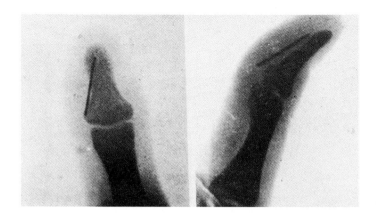

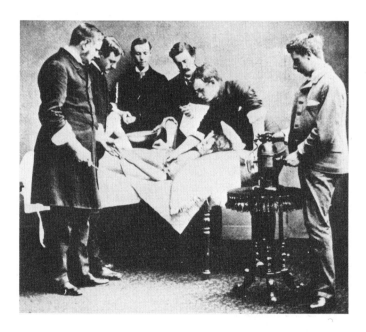

13. Science and Medicine. *Above*, the first X-ray photos used for clinical purposes in Britain, 1896. This was the result of a collaboration between the Professor of Organic Chemistry at University College, who had been present in Germany where X-rays were discovered some two months earlier, and medical colleagues at University College Hospital. The patient had broken a needle in her thumb. *Below*, aseptic surgery, c. 1880. Though surgeons still operated in their germ-ridden street-clothes, the use of carbolic spray was an important development in creating aseptic conditions for surgery.

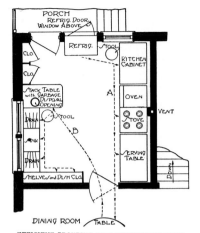

EFFICIENT GROUPING OF KITCHEN EQUIPMENT
A. Preparing route. B. Clearing away route.

BADLY GROUPED KITCHEN EQUIPMENT

THE LABOR-SAVING KITCHEN

Time Study Showing Saving through Correct Grouping of Equipment

Study 1. 1. Walk to storage.
2. Return from storage with small basket of potatoes, and lay on kitchen table.
3. Walk from table to pot-closet for pot.
4. Return from pot-closet to table, on which lay pot.
5. Walk from table to pantry drawer for knife.
6. Return from pantry with knife.
7. Peel potatoes on table surface.
8. Take pot of potatoes in hand and walk to sink.
9. Wash potatoes and fill pot with water.
10. Walk from sink to stove and lay pot on.
11. Walk from stove to table, place refuse in basket.
12. Walk from table to sink with refuse and empty same into garbage pail on floor.
13. Take scrub cloth from sink to table, wipe up same.
14. Return with soiled cloth and knife to sink.
15. Wash cloth, hang up. Wash knife.
16. Walk from sink to pantry drawer to replace knife.
17. Walk from pantry drawer to sink to get basket.
18. Take small basket back to storage.
19. Return from storage.
Time consumed: 5 minutes.

Study 2. 1. Walk to shelf adjacent to sink and get pot.
2. Walk to storage, carrying pot, and fill it with potatoes.
3. Return from storage, laying pot directly on vegetable preparing surface near sink.
4. Pick up knife (from nail above this surface).
5. Pare potatoes directly into pail (soiling no surface).
6. Wash potatoes and fill pot with water.
7. Wash and hang up knife (on nail above sink).
8. Walk with pot and lay on stove.
Time consumed: less than 2 minutes, not counting actual peeling, which would require the same time in each case.

Résumé:

	Time Required	Number of Steps
Study 1	5 minutes	19 steps
2	2 minutes	8 steps

14. Household Engineering – Mrs Christine Frederick was an American exponent of the application of factory-devised methods of scientific management to household tasks. She argued that rational organization of equipment and work processes would alleviate women's domestic burdens. *Above*, her view of well and badly arranged kitchens. *Below*, her time and motion study of a food preparation process in the two types of kitchen.

15. Henry Ford's revolution in automobile production at Detroit, Michigan. *Above,* static assembly of the Model N at the Ford Motor Company Piquette Avenue Factory in 1906. The moving assembly line was introduced in 1913 for the mass production of Model Ts at the Highland Park factory, and streamlined and extended at the River Rouge Factory. *Below,* the Model A final assembly line at the Rouge in 1928.

16. Two examples of attempts to bring science and technology to bear on human reproduction and sexuality. *Left*, an undated poster issued by the British Eugenics Society. In the interwar period, eugenists gave up earlier attempts at legislative reform, and concentrated on publicity campaigns to disseminate their beliefs on who was fit to procreate. *Right*, a slot-machine dispensing condoms in Britain during the 1940s, a time in which soldiers in particular were regularly warned about venereal diseases.

The Department of Commerce lighted over 22,000 miles of airways, operated 74 radio broadcast stations, and 137 radio range beacon stations, and had over 13,000 miles of teletypewriter service principally for handling weather information. [. . .]

Lighter than air craft

Use of lighter than air craft of the Zeppelin type within the continental limits of the United States is out of the question. The Hindenburg has no advantage in speed over railroad trains; is more dependent on weather conditions than transport planes, and from neither technical nor commercial standpoints is adaptable to the type of service rendered by airplanes. On trans-oceanic flights, however, such ships have already set high standards for comfort and safety. Even if equal safety be granted—and this of course remains to be proved in the North Atlantic service—the comfort of ships like the Hindenburg combined with its economy of operation should largely offset the superior speed of transport planes. Moreover, these lighter-than-air craft have a real freight-carrying capacity admirably suited to the transportation of luxury goods. Like the transport plane, material failures are relatively few and, in expert opinion, the success the Germans have had with their Zeppelins is largely due to profound studies of meteorology and air navigation as an art in itself by an extraordinarily well-trained personnel.

Future developments in general

While commercial air transportation, on account of the limits of time and space, is primarily adapted to much longer distances than the average passenger journey, it is still capable of large expansion. Bigger planes and more of them will improve the numerical relationship of air-line employees to passengers carried and cut operating costs per passenger mile. Long-distance first-class mail will be handled in much greater volume, if, as has often been suggested, we have a single first-class postage rate and the mail dispatchers are free to send each letter by whatever method is fastest.

The growth of air express traffic, at present in its infancy, will greatly increase—transportation of perishable commodities, such as fruits, flowers, and seafood, and all first-class freight and express, characterized by high values and low weights per cubic foot, will be handled. Older passenger planes can readily be converted to this service and thus have a better chance of lasting out their physical life in service comparatively free from obsolescence.

These considerations all point to lower airplane fares and the reduction from an average per-mile rate of 12 cents in 1926 to 5.7 cents in 1935 will continue to levels probably not much out of line with those of our extra-fare all-Pullman trains.

As transport flights to some degree represent business lost by railroads, so private flying will largely be at the expense of the private automobile.

In the entire field of passenger transportation, however, definite limits are not yet in sight and these shifts from one form of transportation to another will be less important than the new business created and developed by cheaper, more comfortable, safer and speedier transportation agencies.

The social implications of air transport of the future differ in degree rather than in kind from what we have today. High-speed trains and automobiles have long ago rendered county and state lines meaningless, have widened trade territories, and brought once distant regions close together. Air transport does not change, but intensifies these effects.

The effect of aviation on employment will be beneficial. The construction, operation and maintenance of aircraft, their high obsolescence and comparatively short physical life, are favorable employment factors by any railroad standards. More important is the fact that passenger traffic, unlike freight, can be created. The probability is that far more traffic will be created by aviation than is diverted from other transportation agencies.

Urban transportation

City transportation of passengers, at one time almost altogether handled by electric street railways, has more and more drifted to the motorbus and the private automobile. A few large cities, notably Boston, Chicago, New York, and Philadelphia, have extensive suburban passenger services operated by steam railroads (in some cases with steam locomotives, in others by electric power). Most cities, however, depend principally on trolley cars and motor vehicles.

Facing generally unregulated and destructive competition over most of the past 20 years, the electric street railways have lost about a third of their traffic. Fifteen and one-half billion passengers were carried in the peak year of 1923, and about 10.3 billion in 1935. Like the steam railroads, the street railways are battling to hold what traffic they have left and to regain some of the business they formerly enjoyed.

Like the steam railroads, too, the streetcars now have to meet the demands of a public which requires a good deal more than the bare necessities of transportation. Comfortable lighting, frequency and regularity of schedules, quietness of operation, and greater speed, are being asked of all passenger transportation agencies and the streetcar is in no way an exception.

The future of urban transportation seems unusually obscure. Probably this is because the trends of all urban development—particularly in the larger metropolitan areas—are uncertain.

The National Resources Committee in their 'Studies of Urbanism' are making an exhaustive analysis of mass transit facilities and their effects on the internal development of urban communities and regions. They are considering not only the different forms of urban transportation, the effects which the internal organization of urban communities have on the use of city transit, but also the possible modifications in transit and traffic practices and policies advisable in furthering healthy urban and regional growth. This work when completed should shed a good deal of light on a decidedly dark subject.

Today, however, a typical forecast of the future of the electric street railway is that of the Federal Power Commission in its recent survey of the use of electric power in transportation. The Commission says:

It is dangerous to prophesy, with conditions changing so rapidly, just what effect these developments will have on the electric railway industry. There should be a rebound in its business from the depression lows, with an increase in power consumption, but whether or not the trend will continue upward or begin another decline is unpredictable.

Perhaps, however, a rather surprising lack of competition in this particular field of prophecy should encourage rather than deter a student of transportation. It may still be possible to venture somewhat cautiously where 'angels fear to tread'.

Elevated railways
These are definitely on their way out. Their noise and unsightliness, the inconvenience of climbing to their stations, their devastating effects on property values make any extension of this form of transportation highly improbable.

Some of these roads have already been torn down, and it is probable that we shall have a gradual elimination of others.

Subways—rapid transit systems
The construction of subways depends on concentration and density of population—that is to say, the geography of each individual city. Manhattan, a long, relatively narrow island, where millions of people must be transported in and out of a comparatively small area daily, could not function without subways. Cities where the population is spread over wide areas, as in Detroit or St. Louis, present a totally different problem, making subway construction unlikely.

Like electrification of a steam railroad, it is almost entirely a question of traffic density; and few of our cities, regardless of how large their total population, have enough potential traffic concentrated along any one line to warrant the cost of subway construction.

Subways—street railway
Based on Boston's 40-year experience of routing surface streetcars through a comparatively short subway loop, the suggestion is frequently made that this scheme might be a solution for the traffic problems of many cities where real rapid transit subways, like New York has, are out of the question.

Here, however, is an expensive form of construction which is not high-speed transportation. The Boston streetcars were forced underground by sheer congestion of relatively narrow streets, but the local geography of Boston is admittedly unique. Were some similar plan applied to the ordinary city, the benefits to the individual passenger would probably be relatively unimportant. Instead of stopping at every corner, stations could hardly be spaced more closely than every three or four blocks, necessitating correspondingly longer walks for the individual. Climbing and descending stairs, passing through turnstiles, hunting up a particular spot on a long platform in order to take a car for the desired route, and similar minor delays, would all consume a good deal of the individual passenger's time. Against this, any increased speed would be but a minor offset owing to the relatively short lengths usually discussed for such subways.

Trolley bus
This type of transportation consists of a bus using two overhead wires for an electric circuit, but running on the street like any other vehicle—'a

trackless trolley.' Considering the aesthetic objections to overhead trolley wires and the long effort of most cities to get all wires underground, this type of transportation will probably not be received with general enthusiasm.

Where streetcar tracks have been pulled up and the overhead wiring remains, operation of trolley buses may serve as an intermediate phase of transportation. That most cities, however, would widen a street, permit streetcar tracks to be taken up, and still keep overhead wires, to which, of course, practical objections (in case of fire for instance) as well as aesthetic, may be registered seems improbable. Even less probably would a city permit overhead wires to be strung where none now exist.

Trolley lines

Like the steam railroad, the electric street railway has been peculiarly vulnerable to automotive competition in regions and on lines of light traffic. Twenty-six thousand miles of city and suburban street railway tracks were operated in 1922 and about 20,000 in 1932—a decline of 22 percent. Meantime the operation of 38 percent of interurban track miles was abandoned. The street railway has proven incapable of shifting its lines to meet shifts in business and population within our metropolitan areas. Where a reasonable density of traffic is available, the trolley car has been more successful, and this is particularly true where new and improved types of cars have been put into service.

Strongly influenced by automotive design, the new cars, such as now used in Washington, D.C., for example, provide much more smooth and rapid acceleration and deceleration. Noise and vibration have quite largely been eliminated; seating is more comfortable, illumination, heating, and ventilation improved, weight less, and great improvements have been made in the highly important feature of appearance, both exterior and interior.

The street railway, however, on the whole seems to be a good example of the type of things which can go ahead for a while because they are already here, but which would not be built under today's conditions. Logically it should follow, and actually it has followed, that when the street railway is confronted by the necessity of major changes, such as complete reconstruction or relocation of tracks, the more economical policy is to discontinue operations and substitute automotive transportation. For this reason it seems as though on the whole, in most cities, the busses will rather gradually gain on the street railways.

Busses

Because it does not have to provide its own right-of-way, the bus is well adapted to operate with the utmost economy in light-traffic areas. Busses also can be economically built in a wider range of sizes than streetcars, providing a desirable flexibility of services; the small bus to some degree makes up in speed and frequency of service what it loses in passenger capacity. Mention has also been made of the fact that busses can readily shift their routes with the development of particular sections of the community or the stagnation of others. In periods of temporary traffic interruption occasioned by accidents, fires, broken water mains, and other hazards of metropolitan life, busses may readily be detoured. From the standpoint of the individual passenger, bus stops at the curb are far more convenient than streetcar stops in the middle of wide boulevards full of rapidly driven automobiles. Not only from the disabilities of the street railway but from these definite superiorities of busses, the latter should gradually gain on the 'trolleys.'

Private automobiles

To how great an extent the private automobile figures in urban transportation has never been determined statistically. That it is the dominating factor outside of travel to and from central business districts is plain. To what extent it is used to go downtown and back is largely a question of the local geography of individual cities. An extremely small portion of people working on lower Manhattan Island use private automobiles to or from their offices and, obviously, a large portion of the population of places like Indianapolis and Kansas City drive their own cars to and from work.

Where a reasonably speedy and frequent bus or streetcar service is available, the average automobile owner seems entirely willing to use it. To obviate the difficulties and delays of driving and particularly of parking, to save money in the process and to gain safety and comfort, especially in extremely hot, wet, or cold weather, are worth a good deal to the ordinary businessman or shopper.

General considerations

The most important single factor in urban transportation in the average community is the use of its downtown streets. If they exist largely as storage facilities for parked automobiles, neither bus nor streetcar can supply truly rapid and convenient transportation. If, on the other hand, the downtown streets are generally cleared of all but moving vehicles, the problem of urban mass transportation is well on its way to solution. Certainly any consideration of the greatest good for the greatest number points toward elimination of downtown parking, which would lead to provision for parking facilities on the outskirts of congested districts. Failing substantial progress along these lines, it is by no means inconceivable that private automobiles may be denied access to certain areas, at least at certain times of day. It seems unlikely that people who ride 40 or 50 in a bus, or 75 to 100 in a streetcar, will permanently suffer the delays and inconveniences occasioned by private motorcars occupying equivalent spaces in the streets.

Future social implications of transportation

For two reasons little has been said about the future social implications of transportation.

In the first place we have a great and highly efficient transportation system today, the social effects of which are plain to all of us. This system is constantly being improved in speed, comfort, flexibility, and in lowered cost, but this progress is in degree rather than in kind. We are improving a wonderful machine. We are readjusting the machine to handle a freight traffic consisting more largely of consumer goods and relatively less of capital goods than in the past, and to serve the rapidly increasing passenger traffic of a surprisingly nomadic society. We are not, however, discarding transportation service in favor of something else as, theoretically at least, our current models of contented cows may be rendered obsolete by synthetic milk from a chemical plant.

In the second place the volume and character of transportation, both freight and passenger, depend largely on factors, social and otherwise, quite outside of the transportation industry. This thesis depends on no narrow interpretation such as making the rather obvious point that transportation is completely dependent upon power. Rather let us realize that the exhaustion of mineral resources, or the destruction of timber in a region may leave no freight traffic in that area. A failure of the California citrus fruit crop, a drought in the Kansas wheat belt, a strike closing the Illinois coal mines, the completion of a great new hydroelectric power plant in Tennessee—all such things have a profound temporary and frequently a permanent effect upon the transportation industry which itself cannot take any counteraction. Similarly, shorter hours, higher wages, greater old-age security, and better education will favor increased passenger travel just as the long hours and poverty attendant upon farming submarginal land virtually root people to the soil.

Within the transportation industry itself the social prospects for the future are good. The internal shifts will probably be toward types of service employing relatively more operating personnel per traffic unit than the business of the past has required.

Probably three-quarters of our million railway employees are engaged in a freight service handling 236 billion ton-miles annually: motortrucks, according to the Automobile Manufacturers' Association, employ 2½ million drivers alone to handle 10 or 15 percent as much business. This indicates that motortrucks furnish 20 or 30 times as much employment per ton-mile as do railroads. While the figures are not completely accurate (as previously pointed out) nor the statement of the case entirely fair owing to the somewhat different nature of the traffic, so wide a difference may be liberally discounted without impairing the validity of what it implies.

Similarly, air traffic, which will increase at a higher rate than other forms of transportation, is a liberal employer of high-class personnel.

Changes in freight transportation do not ordinarily affect us directly. Most people have never even seen a pipe line; millions have never heard the whistle of a big ore carrier or a tank steamer. A long freight train may occasionally block us at a grade crossing, or a big truck get in our way on the road. In passenger transportation the contact with our daily life is a good deal more direct. If we live near water we may take a steamer occasionally; more frequently we ride on trains; less frequently we may fly.

One transportation agency, however, is part and parcel of our daily life–the highway and the motor vehicles that use it. It is unnecessary to attempt to catalog the obvious influences of motor transportation ranging from its evil effects upon the hammock business through the many conveniences, the different kinds of usefulness, the broadening recreational opportunities, the recasting of the pattern of farm life, and so on to the vision (or nightmare) of half our population living in trailers. Even the slightest reflection as to how elimination of all motor vehicles would affect us would produce a picture quite as appalling as those drawn of a world without electric power. That the world managed to get along without the 'high line' and the motorcar until quite recently (in 1895 four automobiles were registered in the United States) is perhaps beside the point. From the fact that the ordinary automobile owner will surrender almost any of his possessions and make material sacrifices in his way of living, if only he can hang on to his car, we may infer how largely highway transportation bulks directly in our lives. [. . .]

While not merely volumes but libraries might be written as to the part transportation plays in our civilization, the essential facts are too well known to demand inclusion in a paper of this scope.

Two quite modern and still rapidly developing influences should, however, be noted. First is the improved quality of freight and passenger service, particularly in speed and reliability. The producer, the manufacturer, the merchant, the public utility, or the individual is getting a freight service measured in hours instead of days, and in days instead of weeks, even as compared with 15 or 20 years ago. With this increased speed has come greatly increased reliability, and the influences of both have been profound.

Inventories need only be sufficient for a few weeks, where months were formerly necessary; goods are in transit a few days instead of a matter of weeks; a tremendous amount of capital formerly locked up in these items is now released. Investment of capital in storage facilities may be diminished; the cumulative processes of converting raw materials into finished goods are steadily being speeded up. Carrying less stock, stores become decreasingly vulnerable to changes in styles and to whims of public fancy; with less money tied up in their inventories, manufacturers are more ready to modify their products to suit changing demands.

Improved service affects the transportation industry itself by requiring less equipment.

Granting that speed means less to a distillery aging its products for several years than to an automobile manufacturer who proverbially wants something delivered yesterday that will not be ordered until tomorrow, the effects of better transportation are almost universally felt and of increasing importance. This trend will continue.

Increased speeds in passenger transportation permit much broader and more intensive solicitation and servicing in a business way, and a greatly increased range of recreational and semirecreational travel.

The second great development in transportation in recent years has been the increasing ability of the individual to supply his own freight and passenger service and to some degree render himself independent of common carriers.

The overwhelming bulk of passenger transportation is already an individual matter—the private automobile dominates the field both in business and pleasure travel.

In freight traffic the petroleum pipe lines have a common-carrier status but in the main transport the oil of the companies which own them, or with which they have a community of interest. Even the outside shippers are generally large and represent other units of at least partially integrated oil groups. Oil companies also own and operate large numbers of tank steamers.

Probably about 85 percent of the motortrucks are privately owned and operated, and the total volume of freight handled by trucks is increasing more rapidly than the business handled by railroads or waterways.

This growth of privately owned freight and passenger transportation agencies is chiefly important for its influence on rates. Passenger rates of common carriers cannot stray too far from what motorists think their travel costs them—unsound as it is for a car owner to keep a record of his gasoline purchases and ignore the far more important items of ownership costs such as personal property taxes, insurance to cover half a dozen contingencies, and above all depreciation.

The present freight rate structure will be materially modified through the trucks' influences on pickup and delivery service, packing requirements, more liberal classification of freight, and the level of the charges themselves being based, to an increasing degree, on cost of service. Ordinarily lower freight rates widen trade territories and intensify competition rather than increase the volume of business. More ton-miles may be produced, but with a narrower margin of profit.

The Interstate Commerce Commission must have in its records thousands of pages of testimony and exhibits tending to prove how small are the freight charges on a necktie from Paterson, N.J. to Syracuse, N.Y., and a pound of flour from Franklin, Tenn., to Louisville, Ky. Quite dissimilar testimony pictures the alleged crushing burden laid upon commerce by rates on gravel or coal.

Whether or not lower freight rates mean much on any specific commodity is immaterial; transportation costs in the aggregate are a large element in the costs of our material civilization. Anything reducing this element, provided it allows a fair living to the transportation agencies themselves, is plainly beneficial.

We have a transportation system made up of widely diverse agencies. Some of them are closely related and furnish comparable competitive service—for instance, railroad and highway transportation. Others in the nature of things can be neither competitive nor cooperative—the transport plane and the pipe line are the best examples.

Mississippi River vessels compete with railroads on freight, but not on passenger business. The pipe line has virtually driven the railroad out of hauling crude petroleum, but today offers no competition on other commodities save gasoline. The bulk of freight transportation is performed by common carriers. The privately owned motorcar dominates the passenger field.

While the average man will not be inclined to quarrel with the statement that each form of transportation has its peculiar advantages and that all should be used in a way to permit the greatest utility for each, the practical difficulties in achieving this desirable end are plain.

Are we to say to the truck owner that he may not haul freight between Chicago and the Missouri River because the railroad is a more economical means of transportation? Shall we, as a matter of public policy, deliberately seek to force traffic to one transportation agency because it is necessary to our system of national defense or for other reasons? Shall we turn over the mail, so far as possible, to the air lines and fix their passenger rates in some definite relationship to railroad fares?

Probably we may arbitrarily do a great many things of this sort as temporary or experimental measures even in peace times and more certainly under the stress of war. The difficulties of doing anything of the kind under normal conditions, however, are obvious. Even were all our forms of transportation owned and operated by one agency instead of by hundreds of companies and literally millions of private individuals, the problems involved are entirely too large and far reaching to encourage one's sense of omniscience. Nor are the problems of transportation static—the movement of freight and passengers is in a constant state of flux, shifting with changes in agriculture, industry, trends of population, and all the multifarious factors of our modern civilization.

The difficulties and the magnitude of transportation problems do not preclude the possibility of beneficial and effective coordination and regulation. They are worthy of careful scientific and sympathetic study by the ablest men we have.

It seems clear, however, that whatever regulations we have should be uniform. If we fix rates for one transportation agency, its competitors should not be free to make their own rates. If it is necessary to establish hours of service for the driver of a common carrier motortruck, it is equally necessary in the case of a privately owned truck.

Probably regulation of transportation in the future will be needed more for coordination and the prevention of unfair competition than to bring about reductions in rates.

That some measure of regulation will be applied, even to private, as well as public carriers, seems inevitable.

In the past 15 years we have seen an almost incredible increase in passenger transportation, due to the convenience and low cost of private automobiles. Recently reductions in long-distance telephone rates in Great Britain have greatly increased the gross receipts of the British Post Office, which operates the telephone lines in that country. Today, lightweight, high-speed, streamlined trains and low passenger fares are increasing railroad travel. It is notable that transportation, as it becomes more speedy, regular, frequent, economic, or efficient, creates, like other public services or utilities, an increased demand.

The history of transportation has shown that as a whole it grows at a rate higher than the increase in population. Our transportation service will hold its world leadership; it will continue to develop and adapt itself to the changes in the American scene, and it will become increasingly adequate and responsive to whatever demands are placed upon it.

5

Communications

5.1 T.A.M. Craven and Committee, *Communication by Wire and Wireless*, 1937

[...]

Today the necessity for rapid, efficient, and cheap communications has been thoroughly recognized as a well-established benefit to mankind. Modern communications, while still inadequate in various parts of the world, are a most important adjunct in the economic life of nations and individuals, and they have borne and will continue to bear a most important influence upon the social and economic welfare of the entire world. A single illustration will serve to demonstrate this fact when one considers a comparison between the difficulties of government in the United States during the early days of the Republic and the present day. If the 130,000,000 people in the United States today utilized the methods of communication of 1776, no one can state that our present form of government, our economic structure, and our social habits would be as we know them today. Looking into the future, with the advent of radiobroadcasting, one can visualize further improvements and benefits in the art of self-government, because even though broadcasting is today only 15 years old, its effect upon the life of the country has been most marked because of its inherent ability to provide for instantaneous mass communication over tremendous distances, thus welding the people of the country into a unit unparalleled in history.

When one adds to aural broadcasting the ability to see and to record permanently what one has seen and heard, there will have been attained a perfection in communications which will certainly have the most profound effect upon our social and economic life.

Broadly speaking, the most important technological developments which will enhance the effect of communications upon the future public life are those which will increase the speed and availability of inexpensive communications. These are now in the process of development, in the form of facsimile, television, and methods to increase availability of channels. The first enables the instantaneous transmission of permanent records, photographs, the printed page, and other signs; the second permits the transmission and reception of objects, and the third increases the number of voice and record channels and cheapens the cost of communications.

Source: T. A. M. Craven and Committee, 'Communication by Wire and Wireless', in National Resources Committee, *Technological Trends and National Policy including the Social Implications of New Inventions*, US Government Printing Office, 1937, chapter 4.

Of course there are hundreds of individual inventions which will tend to improve the details of facsimile, television, and multiplex operation, but the principal developments are those which now make it possible to apply to public service facsimile or television either through the use of land line or through the use of radio, either for person to person contact accompanied by voice, or for mass communication. Probably the most important developments in the application of these broad systems of communications will be the organization of facilities to make their use by the public possible and more easily available. In radio there will be necessary many inventions to increase availability of television and facsimile to the public, since at the present time there is an inherent natural limitation in the use of the radio spectrum which has yet to be conquered by man's ingenuity.

There will be required in the future a balance between the unlimited possibilities of facsimile and television as against the limitations imposed by natural, economic, and social circumstances. Nevertheless, one can safely state that ability to have one's newspaper printed in the home, ability to see and hear news in the making, ability to transmit quickly from one point to another a written document, and the ability to see and hear the person with whom one desires to communicate, even though he may be separated by thousands of miles, must have a marked effect upon the daily life of anyone living in this modern age. That this may affect the social habits and daily routine of an individual family or that it will affect the economic welfare of the Nation as a whole, cannot be overlooked.

Adjustments will have to be made to accommodate the rapid changes in the panorama of the future as compared with that of today. With the ability to see and hear from persons at a distance when one is traveling by airplane, automobile, or steamship, as well as when one remains at home, or as one conducts business in an office or works in a factory, it appears inevitable that the mental processes of the future must be such as to produce an entirely different outlook than exists under circumstances when one's vision, horizon, and social contact are limited.

It also may be expected that these new developments will speed up ordinary life and business, and will affect certain existing industries, such as the motion picture, the newspaper, advertising, and the existing telegraph, telephone, and radio systems of the country. The effect of these developments upon the industry and their consequent effect upon commercial activities in every walk of business life, requires modifications of economic views which exist today.

In education, the application of these modern methods of communication may well effect a complete change in methods of educating not only the child and the adult but also the entire public. The new communication development may also revolutionize the present school systems of the country.

Whether or not these new developments will permit more leisure and greater profits will depend entirely upon the control which the public places upon such new developments, particularly as to the organization and methods of making these wonderful facilities available for use by the public at the cheapest cost. In consideration of this factor alone, one must take into account the economic limitations which will be involved in the rapid obsolescence of present-day communication facilities.

It is considered by many that the application of these modern communication facilities will not result in a decrease in employment, but rather in an increase in employment. However, such employment will tend toward those who are qualified scientifically rather than those who are qualified manually. The person of the future may, as a result of propinquity with the everyday modern communication developments, become a far better educated thinker than the average person of today. [. . .]

Telephony: wire and radio

The importance attained by telephony in meeting the industrial, commercial, agricultural, financial, economic, social, and other needs of our present-day civilization rests largely upon the fact that it has made possible complete two-way intercommunication between practically all the telephones of this country and between over 90 percent of all the telephones in the world, with a relatively short delay in the establishment of a desired connection, with ample volume of transmission, and with such a high quality of reproduction of the transmitted voice that the speech characteristics of the talker are readily recognized. Further, the speed of transmission, varying with the circuit facilities employed from 10,000 miles per second to almost its theoretical limit of 186,000 miles per second (the velocity of light in vacuo), is so great that ordinary conversations can be maintained almost as they would be face to face without even noticing the fact that the transmission is not instantaneous. [. . .]

Changed Population Characteristics—When the telephone was invented the population of the United States was overwhelmingly rural but it is now only 44 percent rural. The corporate limits of such cities have in many cases been practically ignored in the development of suburban areas, many of which have themselves grown into separately incorporated towns and cities, the whole often forming a continuous urban community. In 1930 the United States census reported 93 cities as having populations of 100,000 or more with an aggregate population of over 36,000,000. It also recognized 96 'metropolitan districts' composed of one or more central cities, together with the surrounding towns and villages, thus forming one area having more or less common social, economic industrial, and financial interests. In 1930, these 96 metropolitan districts had a combined population of almost 55,000,000, 5,000,000 more than the entire United States population when the telephone was invented.

Telephone development, until the depression, much more than kept up with the increase in population. It has been designed to the end that telephone service will be offered to every community, large or small, in accordance with the local requirements, the equipment naturally varying according to the size and telephone development of the area served.

At one extreme we find many rural areas served by multiparty 'farm' lines equipped with local battery telephones and magneto ringers and signals, the lines in most cases being connected for switching service to the next nearest exchange.

At the other extreme is the New York-northeastern New Jersey metropolitan district, a description of which follows:

More than a sixth of all the telephones are concentrated in a section which composes only about one twelve-hundredth of the land area of the continental United States.

This section, as defined by the United States Census Bureau, is the New York-northeastern New Jersey metropolitan district, generally called the New York metropolitan area. It includes large sections of those two States and a small part of southwestern Connecticut, comprising a total of 2,514 square miles, a territory twice the size of Rhode Island. An exact circle enclosing such an area would have a diameter of about 57 miles. In its actual form, and using the Borough of Manhattan, New York City, as the center, the longest axis is approximately 95 miles and the shortest 40 miles.

Within this area are approximately 300 incorporated communities. Among them, in addition to New York City with its 5 boroughs and population of 7,000,000, are Newark with about 450,000, Jersey City with more than 300,000, 3 other cities with well over 100,000, 8 with over 50,000, 15 with over 25,000, and more than a score with over 10,000.

The area's aggregate resident population—10,901,424 according to the 1930 census—exceeds that of any State except New York. Computed on the basis of that census, it contributes nearly an eleventh of the population of the entire country.

Approximately a sixth of the average daily telephone conversations originate from the telephones in this area * * *. The area's telephones which serve these myriad requirements numbered as of July 1, 1934, about 2,255,000, of which 1,478,000, or nearly two-thirds were in New York City. Manhattan alone had about 817,000 * * *. The average daily telephone conversations during the first half of 1934 totaled approximately 9,183,000, or about 382,625 an hour.

Subscribers' telephones in the general metropolitan service area, which includes some sections not within the metropolitan census district, are grouped under 478 central office designations, and are served by 378 central offices, an addition of 138 since the war. New York City, with 160 central offices, has 192 designations in use, as many as the aggregate in the 3 next largest cities of the United States. The designations as a whole exceed those in the 10 largest cities other than New York.

Employed within the area in telephone service are approximately 48,000 men and women * * *. About 60 percent are women. About 35 percent of the employees are engaged in plant work, and about 30 percent are operators.

About 1,210,500 telephones, or more than half of the area's total, are served from dial central offices, of which there are 129 * * *. The dial system is regarded as providing special advantages in serving a cosmopolitan population.[1]

From the above it will be seen that there have been developed in the United States local telephone systems for meeting varying population requirements.

Toll service—The first period, up to 1900, marks much substantial progress in extending the range of commercial transmission beyond local areas. Beginning with the 2 miles between Boston and Cambridge in 1876 and from Boston to Providence, a distance of 45 miles in 1880, the range has been gradually extended. In 1892 commercial service was offered between New York and Chicago, a distance of 900 miles. This marked the economic distance limit until the next important development, 'inductive loading', became available.

Inductive loading—The principle underlying the inductive loading of telephone circuits was patented in 1900. The purpose of such loading was to compensate for the effects attributable to electrostatic capacity between the wires of a circuit, thus securing a reduction in attenuation, and at the same time equalizing the attenuation, thereby greatly extending the distance range.

Loading was first applied to open wire circuits including the two 'side' circuits and the 'phantom' circuit derived from two pairs of conductors. This made it possible in 1911 to extend service to 2,100 miles (New York to Denver) and 2 years later to 2,600 miles (New York to Salt Lake City).

Loading was next applied to cable conductors for which the normal attenuation was far greater than for open wire circuits. [. . .]

The first application of loading to underground toll cable circuits was made in 1906 between New York and Philadelphia, a distance of 90 miles. This was extended in both directions to Boston and to Washington, a distance of 455 miles, the whole being completed in 1913. Designed before the advent of the telephone repeater, the cable was made up in part of heavy gage conductors (No. 10 American wire gage) and a heavy type of loading to furnish commercial service between the termini. Some sections of this cable are still in use.

The advent of the repeater and associated developments practically solved the problem of long-distance transmission by toll cables and removed all distance limitations. [. . .] The development of carrier telephone and telegraph systems has already materially increased the total circuit or channel mileage available for use. [. . .]

Toll service is handled in various ways, depending upon various factors. In the case of very large toll centers separate offices have been established for handling this important service. Quoting from the article previously referred to:

> The long-distance office in New York itself is a battery of offices which compose the largest long-distance center in the world. It is the focal point of many important cities in the United States. Through it flashes the traffic which is handled over the radio-telephone circuits to Europe and other overseas points. It is the principal control point for the great radio broadcast chains. It houses the largest teletypewriter exchange. All private wires from New York to other cities, whether telephone, teletypewriter, or telegraph, are brought through this building.
>
> From the toll and long-distance offices of the city a vast network of circuits fans out into the surrounding territory, for it is the regional center for a large section of the Northeastern States. Crossing from Manhattan under its surrounding waterways are 46 toll and long-distance cables, with nearly 18,000 circuits. One of many notable engineering features is the permanent subway cable crossing under the Harlem River, between Manhattan and the Bronx. Through it pass not

only the toll and long-distance highways between New York City and New England points, and toll routes to northern suburban sections, but also the paths through which the greater part of the calls between the country at large and the Northeastern States are dispatched.

General toll switching plan—The removal of distance limitations from toll service through the advent of the repeater and associated developments made it possible to plan a systematic plant lay-out for rendering such service between any two telephones in the United States with a minimum of delay in the establishment of connections, with an adequate volume and quality of transmission, and with very material overall economies.

In brief the plan consists in providing a limited number of 'regional centers' (eight at present) each directly serving one of the regions into which the country is subdivided. The places selected for regional centers, strategically located to serve the country as a whole, were New York, Atlanta, Chicago, St. Louis, Dallas, Denver, Los Angeles, and San Francisco. Each of these was planned to be interconnectd with every other regional center by a group of direct circuits thus forming the basis of a country-wide network.

Next were selected some 140 'primary outlets' each of which was planned to have direct circuits to one or more of the regional centers and direct circuits to all toll centers in the area for which it is the primary outlet. Each of some 2,500 toll centers in turn was planned to have direct circuits to all 'tributary' outlets. In addition there have been provided a number of secondary outlets and secondary switching points furnishing respectively alternate routes for toll centers for reaching regional centers and providing more direct routes, thus reducing back-haul for intra-area business. In addition the plan calls for direct circuits between any other points which may be justified by the amount of traffic to be handled.

As a result of careful selection of the regional centers, primary outlets, and toll centers and provisions for direct circuit groups between other important points 80 percent of long-haul toll calls are now handled without any intermediate switching and 17 percent with one intermediate switch. [. . .]

Relation of telephony to broadcasting—Radio broadcasting as organized today depends almost completely upon wire facilities. The following description is contained in a pamphlet[2] on this subject by an officer[3] of one of the broadcasting companies.

It is to the telephone, not to radio, that we owe the development of the equipment whereby speech and music are made available for broadcasting.

More than this, it is the telephone wire, not radio, which carries programs the length and breadth of the country. John Smith, in San Francisco, listens on a Sunday afternoon to the New Philharmonic orchestra playing in Carnegie Hall. For 3,200 miles the telephone wire carries the program so faithfully that scarcely an overtone is lost; for perhaps 15 miles it travels by radio to enter John Smith's house. And then he wonders at the marvels of radio.

But what about programs from overseas? Here indeed wireless telephony steps in, but not broadcasting in the ordinary sense. The program from London is telephoned across the Atlantic by radio, but on frequencies entirely outside of the broadcast band. [. . .]

Transoceanic telephony—Transoceanic telephone service, utilizing radio to span the Atlantic, was opened to the public on January 7, 1927. Low-frequency (long-wave) stations were used, which were supplemented by high-frequency (short-wave) stations on June 27, 1928. This combination of low and high frequencies provided a considerable increase in reliability and continuity of service. Intercontinental and overseas commercial telephony has steadily expanded from that time until today more than 60 countries may be reached by telephone subscribers in the United States. Some of the countries which are directly connected by radiotelephone circuits with the continental United States are England, Bermuda, Bahamas, Puerto Rico, Santo Domingo, Jamaica, several Central American nations, Peru, Argentina, Brazil, Venezuela, Colombia, Hawaii, Japan, and the Philippines.

There exists certain characteristics inherent in international radio communication which are mentioned here as being of possible interest. It is obvious that differences of time in various parts of the world will affect operating plans and will control the flow of traffic. For example, on circuits between New York and Europe, the greatest daily activity occurs in the forenoon (New York) during the business hours which are common to both regions. On international telephone circuits, difficulties in the use of different languages must be overcome. A further complication arises when circuits are interrupted or delayed because of interference from foreign radio stations suddenly operating on or near the frequency of an established circuit. Matters of this kind require international correspondence, explanation, and agreement in order to restore normal conditions. In conclusion, it may be mentioned that the volume of international message traffic has been shown to vary closely in direct proportion to the volume of export and import trade between the United States and other nations. Also during the principal holidays each year and in periods of international crises the number of international telephone calls have shown a substantial gain.

Mass communication

Broadcasting—aural

The development of broadcasting is very closely associated with the economic, political, and social history of the past 15 years. The system, methods of operation, and means of financing operations as they exist in the United States today should be considered as the almost inevitable result of the growth of an instrumentality for directly serving the public in a democracy where initiative and freedom of speech are fundamental. [. . .]

Broadcasting is today an integral part of the everyday life of most people in the United States. It brings to the fireside finer entertainment than has heretofore been available to the average individual. This entertainment includes comedy, drama, popular music, and concert music. The gaining interest in classical musical programs is evidenced by the hearty response to the Sunday evening classical hours. Sports have an important place on the program schedules of most stations, particularly during the baseball and football seasons.

An important function of broadcasting is, however, the conveying of direct information to the listener. This includes news broadcasts, weather reports, and storm warnings which are of major importance in certain sections, and market and livestock quotations which are an aid to those interested. Broadcasts by public health authorities have rendered notable assistance in preventing the spread of disease in times of crisis such as that caused by the recent widespread floods in the eastern part of the United States. A notable service which may be classified as direct information is the discussion of current topics by prominent individuals in the fields of government, economics, and sociology which helps to acquaint the average individual with the numerous problems incident to modern civilization and assists him in arriving at better conclusions relative thereto. In this respect it has the effect of clarifying the thought of people on current topics and speeding their decisions in national problems. It is possible today to present to a nation within a few minutes through the medium of broadcasting information and discussions which would have been utterly impossible 15 years ago. This fact has a very striking effect upon the mobility of thought and opinion.

The radio with its increasingly permanent place in the home has a unifying effect within that home and it is thought by many that it may be responsible, to some extent, for counteracting the effect upon American home life which has been produced by the automobile.

Broadcasting, with its direct personal appeal, its easy and ingratiating entrance into the home, is in short the most effective and can be the most formidable means of mass communication which man has yet had the privilege of using. [. . .]

As the intensive growth of broadcasting has coincided with the growth of sound movies, it is difficult to evaluate separately its effects. Many of the musical activities which were previously confined to the concert hall have been transferred to the motion picture and radio studios. This has helped to make tremendously popular outstanding members of the musical world. It has raised the taste of the public in musical performers and in so doing adversely affected the small itinerant musical organizations which were known 15 to 20 years ago. In spite of this and the decrease in demand for musical individuals and organizations, since the advent of broadcasting and sound pictures, it is believed that there is an increased interest in the production and enjoyment of music by the amateur musician and music lover. [. . .]

A phase of broadcasting which has widespread effect and is potentially an important factor in the creation and continuation of international good will is international broadcasting. Many of these stations broadcast musical programs, news bulletins, talks, etc., in languages other than the native language for the reception of foreign listeners on frequencies between 6,000 and 25,000 kilocycles. Needless to say, news digests, speeches, etc., which are colored to present a pleasing picture of the Government or other organization broadcasting, will have some effect upon the opinions of listeners in foreign countries.

An increasing amount of effort is being expended in perfecting the mechanism of rebroadcasting, and frequently European programs are presented to American listeners with little impairment of quality due to the long distance over which they have traveled. The reception of programs direct from foreign broadcast stations and the rebroadcasting by United States stations of similar programs are important factors in fostering international brotherhood and good will, and through the development of better receiving equipment, increases in transmitter power and the use of directional antennas, it will be possible to continually improve the quality of this reception.

Broadcasting—visual

[. . .]

The transmission of television images through the medium of wire or radio circuits has been an accomplished fact for at least 10 years. The first pictures were crude, the reproduction imperfect when compared with the modern motion picture and it has been felt by the industries that no system of television would be commercially feasible or receive any measure of public acceptance which could not transmit pictures of sufficient size to be readily usable in the home and of sufficient definition to compare favorably with the present motion picture. Research has centered around these two important factors.

The first systems made use of mechanical means of picking up and reproducing television pictures. What appeared to be a limitation to this means of pickup and reproduction was soon found and, while mechanical methods have not been abandoned, attention was turned to wholly electrical systems. Recent developments in the United States and abroad indicate that a fairly satisfactory picture, approximately 6 to 8 inches square, can now be transmitted by wholly electrical systems, making use of the very high frequencies previously mentioned as a transmission medium.

The system of pickup, transmission, reception, and reproduction required for television is necessarily complex. There are many different systems and many phases of the subject being studied by the various laboratories of the world. It is desirable that before any system of transmission be standardized for use in a country that the organization doing the standardizing, whether it be commercial or governmental, be satisfied that the system under consideration is the best available, that it is adaptable to continual improvement without rendering existing equipment obsolete and that all organizations wishing to transmit television signals will employ the standard system. Television will be a reality in the United States when it appears that a system has been evolved which meets these requirements and that there is a sufficient public interest and support to warrant the establishment of stations to broadcast television programs. One of the limitations which exists today in providing a Nationwide broadcasting service in the United States is the lack of available channels to accommodate television because each such station requires a very large portion of the radio spectrum; for example, 600 times that required by the ordinary aural broadcasting station.

Another limitation lies in the apparent inefficacy of the ultra high frequencies (where space can more easily be provided) for long-distance transmission and hence, there is some grave doubt as to whether television of high quality can be provided for rural areas in this country at a reasonable cost.

There is also some doubt as to whether the low frequencies which are already being used by existing services other than television will be suitable for rendering adequate television service to rural areas even though it be television of low definition. In any event, if rural areas were to be given low definition television and urban areas high definition television, it is certain there would exist economic and other problems in the production of two types of receivers and a certain amount of discrimination. Thus it appears necessary to concentrate television development on means which will enable the occupancy of smaller space in the ether, cheaper costs and methods enabling the standardization of transmission for both urban and rural areas.

The British Broadcasting Corporation is at present equipping studios and installing transmitting equipment in a wing of the old Alexandra Palace in London and it is anticipated that service will be commenced in the summer or fall of 1936 with two stations, each operating a total of 2 or 3 hours per day, one station using high definition electrical system of transmission and the other employing a mechanical system. It is expected that eventually the station employing the mechanical system will be converted to the use of the electrical system. The British Broadcasting Corporation anticipates an expenditure of approximately $900,000 for the development of this service during 1936. A committee of the House of Commons which investigated and studied the potential use of television in England recommended that before any system of transmission and reception is standardized the postmaster general and the board of governors of the British Broadcasting Corporation require the industries which have been responsible for the development of the various systems to agree upon a standardized system and form a patent pool whereby all interested members of the industry could manufacture and sell equipment under patents owned by other members.[4]

The development of receivers for television has progressed to the point where it is stated by several manufacturers that, should a system of transmission and reception be standardized and public acceptance of television warrant quantity production of receivers, they could be marketed at a cost comparable with that of the home refrigerator. Such a receiver would include provision for the reception of the sound associated with the television program. [. . .]

Who is there today who can predict with any degree of accuracy the effect on our home life and our business life of this new communication facility. It is possible today to sit in one's home and listen to voice and music from the far corners of the earth. In the future, this aural intelligence may be supplemented by another appeal to the senses; namely, the ability actually to see what is going on at some remote point, as well as to hear it. Recent tests in this country and abroad have demonstrated this possibility to be entirely feasible, and it is only a matter of refinement in development, reduction in costs, and providing and organizing adequate facilities to extend the available service from a few miles to many thousands of miles.

Color television is already a laboratory accomplishment. It, too, may become practical before long. Developments have already been started in three-dimensional sight and sound and, if we consider past progress in this field, is it too much to expect that a future generation of Americans will be able to sit at their firesides and see reproduced before them in actual colors and in three dimensions, both visually and acoustically, scenes which are being instantaneously transmitted from the interior of some forest, accompanied with all the fragrant odors of nature, and eventually the addition of a vicarious, tactual sensation. [. . .]

Conclusion

Perhaps no industry in the world has had such a rapid growth and such remarkable development as has taken place in communications. We must content ourselves with a study of past progress in this field, and by drawing certain imaginary graphs of past progress, attempt to project them into the future in such a way that certain trends of themselves become apparent.

Even such an elementary attempt becomes difficult when we consider the tremendous implications which underlie the balance between engineering development and the social and economic trends in communication. For example, in the telegraph industry, does the future hold in store a Government monopoly, as is the case in most European countries, a commercial monopoly as many of the leaders in the commercial communications field desire, or controlled competition which would have both some of the advantages and some of the disadvantages of each of the other systems?

In the engineering field, it is quite apparent that development is taking place very rapidly.

One important trend is the great increase in the use of communications for purposes which 10 years ago had not been even considered. For example, the tremendous growth in two-way police systems; the dependence of the aviation industry on radio as a vital part of that branch of transportation; the growth in the use of radio for such miscellaneous services as geophysical exploration parties, coastal harbor radio-telephone service to fleets of fishing vessels and tugboats, harbor communication to fireboats and the great number of incidental services using general experimental frequencies for scientific development.

Probably the most significant trend, however, is the relative imminence of television. Since 1929, television as a scientific tool has been in a rapid process of development in many large and small laboratories in this country. From time to time predictions have been made that 'it is just around the corner', and the particular corner usually referred to was an engineering one. A number of laboratories in this country have now developed the technical phases of this art to the point where it can safely be said that, although many, many technical problems remain to be solved, it is, nevertheless, possible to transmit over a local area of 10 to 20 miles radius fairly good pictures having the clarity and details of the average home moving picture.

The next corner to be turned, however, is an economic rather than an engineering one, and it can be stated briefly in one short question 'Who is to pay for television?' Will the public accept a television service based upon a continuance of the present system of commercial aural broadcasting and its extension into television? Will a 'looker-in' be willing to sit in a darkened living-room at home intently peering into the screen of his television receiver?

It is believed that the greatest service which communications can do in the future will be to provide extensions into the hitherto remote and inaccessible places whereby people who formerly had no means of communication can be connected with the communication arteries of the world. Tremendous progress has been made during the last decade in this direction and, undoubtedly, tremendous progress will take place in the future. The other great forward step in world civilization which can be made is in the effective use of communications, both telegraph and telephone by wire, but more especially by radio, in the development of understanding, mutual respect and tolerance among the nations of the world. Much has been done along these lines in the past and a great deal more is expected in the future. [. . .]

Notes

1   Excerpts from an article by Kirtland A Wilson, Transmitter, Chesapeake & Potomac Telephone Co., March 1935.
2   Broadcasting: a New Industry, Harvard Alumni Bulletin, Dec. 18, 1930.
3   Broadcasting Network Service , Mar. 1, 1934, p. 10. The officer referred to was H. A. Bellows, formerly vice president of Columbia Broadcasting Co., president of Northwestern Broadcasting Co.
4   Report of the broadcasting committee, 1935, presented by the postmaster general to Parliament by command of His Majesty, February 1936. Report of the television committee, presented by the postmaster general to Parliament by command of His Majesty, January 1935.

6

Food

6.1 Siegfried Giedion, *Frozen Food*, 1948

Frozen foods

The time of full mechanization brings yet another penetration of organic substance: we recognize that it is one thing to keep an organic substance lingering in the neighborhood of freezing point and another to freeze it swiftly by use of low temperatures. A slow descent to freezing point—32° Fahrenheit—bursts the cells of plants and animals. In quick freezing these cells remain intact and hold their flavor like wine corked in bottles.

While wintering in Labrador, Clarence Birdseye, as is well known, observed that the flesh of fish and reindeer congealed rapidly in the Arctic air. When the Eskimos returned months later, it was as fresh as the day when killed. Birdseye translated this into mechanical terms by bringing food to freezing point between metal plates. Soon after this process was patented in 1925, its commercial application began. In 1928 the first food-stuffs processed by this method reached the market. Their consumption grew by leaps—from 39 million pound-cartons in 1934 to 600 million in 1944.[1]

The sculptor Brancusi heard the dictum in the far East that fruit should not be eaten over thirty miles from the place where it grew. Quick freezing will perhaps help to supersede this wisdom, for it enables fruit to be plucked when fully ripe. 'Quick freezing starts at the moment of maximum palatability.'

Similarly with sea food. The catch is frozen as soon as it is lifted aboard the trawlers. Not even the entrails need be removed. In New York we have eaten crab from the Pacific Ocean tasting as if it had just been taken from the sea—far fresher, certainly, than when it has passed through the local markets in the traditional way, or has been extracted from the can.

The economic necessity of shipping livestock from the Great Plains to the Chicago or Kansas City slaughterhouses can in principle be dispensed with. Cattle can be processed on the farm.

What implications are in store?

The economic advantages are evident. Quick freezing affords protection against waste. 'Through refrigeration the farmer can preserve his entire crop and can now obtain full realization of his investment.[2]

Even more important seem to us the latent social potentialities. Quick freezing may be a way to attain better equilibrium in the matter of mass production and monopoly. Rightly used, it should help decentralization.

Source: Siegfried Giedion, *Mechanization Takes Command, A Contribution to Anonymous History*, London: Oxford University Press, 1948, Norton Reprint, 1969, pp. 604–606.

It gives the small farmer a new chance to have his produce compete with the giant farms. He may install a freezer on the farm, as described in Boyden Sparks's small book, based on the first ventures in this direction, *Zero Storage in Your Home* (New York, 1944), or else a locker plant[3] may be operated by the community and made available to all on a co-operative basis—as has already been done in recently developed places. At an early date, in 1936, the farmers of the Tennessee Valley Authority region set up a co-operative freezing plant. Perhaps it will prove the means for an awakening of community interest. The locker plant may be part of the small civic center that must be planned today for every community of a few thousand people. Whether this is the course to be followed or whether the locker plants will become part of a gigantic concern extending from the Atlantic to the Pacific depends finally upon the will of the citizens.

What will be the influence of quick freezing upon city dwellers? Here, too, various outcomes are possible. We shall do no more than touch upon the two extremes.

In an American kitchen, fully pictured in *Life Magazine*, stands a heavy table, a cutting block as in the butcher's store. A butcher's block amid full mechanization? The architect, Fordyce, has indicated a white enameled receptacle, the quick-freezing chest, intended to preserve large sections of meat bought whole and requiring a block in order to be cut correctly.

The time of full mechanization brings back the possible laying-in of a store of meat or other foodstuffs, even by city dwellers. In 1945 fashionable New York apartment houses were to install basement locker plants with at least one locker for each tenant.

Provisions set aside almost on the medieval pattern; contact with natural materials, not cans; craftsmen-like pleasure in spontaneous culinary preparation. These things are possible too.

At the other extreme:

> Meats will be cooked in ton-sized batches under the direction of world-famous chefs and packaged in containers. Then, one minute before dinner time, the housewife will place the pre-cooked frozen meal into a special electronic oven. This oven will employ high frequency radio waves which penetrate all foods equally . . . in a few seconds a bell will ring and the whole dinner will pop up like a piece of toast.

Such is the picture that some writers have painted at the end of World War II to tempt the palate of the American public.

Is not the infra-red oven progress? The housewife need not waste a moment opening cans and waiting for the food to warm. All is done instantaneously. She does not even wash dishes, for the plastic container is thrown away.

In 1945 a number of frozen-food centers designed for 'self-service,' where the cartons are stacked in white enamel chests, were seen in New York and its suburbs. Will the frozen-food centers contain the fresh raw materials of cooking, or will the dominion of the tin can be further extended in the guise of ready-cooked, ready-made frozen foods? Will the assembly-line steak triumph, or will there be a return to spontaneous cooking in the home? As with the fate of rural locker plants, everything depends on the attitude of the consumer.

Notes

1   George F. Taubeneck, *Great Day Coming!*, Detroit, 1944. p. 185.
2   Ibid.
3   'A frozen-food locker plant,' as officially defined, is 'a term applied to a modern low temperature food storage. The services of such a plant include processing, preparation, and freezing. The principal components of a frozen-food locker plant are:
    1   A chilling and aging room at 36–8° F., where fresh killed meat is chilled and aged and other products are chilled, prior to processing and preparation for freezing.
    2   A processing room where meats are cut to order. Meats, fruits, and vegetables are packaged prior to freezing.
    3   A quick-freezing cabinet is used to produce a completely frozen product.
    4   A locker room containing several hundred separate lockers for rental to individuals. Temperature 0° F.
Report of the *Task Committee, War Production Board*, July 1944, reprinted in Taubeneck, op.cit. p. 375.
4   S.S. Block, 'New Foods to Tempt Your Palate,' in *Science Digest*, New York, Oct. 1944.

6.2   Lord John Boyd-Orr, *The Food Problem*, 1950

The hunger of two thirds of the people is a serious economic and political threat to the rest. A summary of the problem and an approach that may alleviate it.

The food problem confronts the world with two dangers. One is the political danger of hunger. A lifetime of malnutrition and actual hunger is the lot of at least two thirds of mankind. Hungry people who believe that an abundant supply of food is possible will overthrow any government that does not make it available. The upsurge in Asia, the most important political event in the world today, is fundamentally a revolt against hunger and poverty.

Side by side with the political menace of hunger there exists in the now small world an economic danger, arising from diametrically opposite causes. This economic difficulty is due to the ease with which food production can be increased with the help of modern technology. Certain countries, the U.S. and Canada in particular, are embarrassed by surpluses of food. To prevent a slump in agriculture, the U.S. Government has taken billions of dollars worth of food out of the world market and is now embarked upon a program of restricting production. Communists are hopeful that the capitalist system will break down because it cannot be adjusted to carry the great wealth produced by modern technology.

In a well-ordered world the danger of revolt against hunger and the threat of food surpluses would cancel each other out. The necessary adjustment could well begin with a world food policy based upon human needs. It will not be easy, but the effort must begin soon.

Source: Lord John Boyd-Orr, 'The food problem', *Scientific American*, vol. 183, August 1950, pp. 11–15 [3 bar diagrams have been omitted from this source].

The gravity of the food problem is such that some observers have arrived at the hopeless conclusion that the 19th century English economist Thomas Malthus was right, i.e., that population tends to increase faster than the supply of food, and that part of the population inevitably has less food than it needs. Certainly the population of the world is increasing at an accelerating rate. At the beginning of the 19th century it was estimated at a little over 900 million. At the outbreak of World War II it had reached about 2,000 million. It is now about 2,250 million and is increasing at the rate of more than one per cent or about 22 million a year. At this rate the world population will reach between 3,000 and 4,000 million during the lifetime of our children.

If the kind of effort being made by the World Health Organization succeeds in eliminating preventable diseases, the rate of increase in population will be much greater than one per cent. The life expectancy at birth of half the population of the world is only 30 to 40 years, compared with 65 to 70 years in countries where modern preventive medicine has been applied. Thus the first efforts to relieve destitution in the overpopulated countries raise the prospect of a calamitous 'explosion of population,' unless measures are carried out concurrently to provide food and the other primary necessities of life for the increased numbers.

The anticipated increase in population must be considered in the light of pressure of population on the land. In the 19th century the most rapid increase in population was in Europe. Despite the emigration of millions to America and Australasia the population increased by more than 100 per cent, compared with a little more than 60 per cent in Asia. Cheap food from the virgin soils of the Western and Southern Hemispheres, however, successfully met this increase in demand. For a time Europe enjoyed the illusion that there was no end to the new arable land. But the last new land which could be easily brought under the plow was broken during World War I. Though some old pastures were plowed up during World War II, practically no land which had not been previously cultivated was added to the arable land of the world. We have exploited the last of the world's virgin soil. Today the very serious evil of erosion, dramatized by Fairfield Osborn in his book *Our Plundered Planet*, confronts us with the prospect of an absolute decline in arable land.

A rapidly rising tide of world population on a diminishing acreage of arable land—this is the food problem. Some believe that the only solution is birth control. Birth control propaganda might help, but it will not act quickly enough to equate population and food supply in our day. Speaking for the neo-Malthusians, William Vogt suggests in his book *Road to Survival* that the road to survival for some of us lies in stopping preventive medicine, assistance in food and efforts to increase food production in the overpopulated regions until disease and hunger have reduced their populations to the numbers the land will support.

There is another view of the world food problem which conflicts with this gloomy outlook on the future. Many observers believe that with modern agricultural and engineering technology the only practical limit to food production is the amount of capital and labor devoted to it. Land damaged by soil erosion can be, and indeed is being, reconditioned not only in the U.S. but in North Africa and Asia Minor, where once-fertile land is now desert. It is now difficult and costly to break in new land, as the British East African groundnut scheme has shown, but it is being done. Frequently it will involve the application of other than strictly agricultural sciences. The elimination of the tsetse fly on the African veld, for example, offers the prospect of opening up a vast new area of grazing land on a continent which is not only an agricultural frontier but also is not even fully explored.

It is claimed that the land presently under cultivation could support twice the present world population, if it were made to yield to the full capacity possible by modern technology. The most frequently cited example of overpopulation is India, with "its three mouths to two rice bowls" and its population increase of four million a year. Yet the task of doubling the food output of India has now been thoroughly investigated; the means are at hand, and the engineering program to this end is already underway. At present the Indian yield per acre of rice and wheat is little more than a third of that in Japan. This is because the land is starved for humus, fertilizers and water. The humus and fertilizers can be provided, and plenty of water falls on the land. India is lacking only in storage and irrigation facilities to use the rainfall to the best advantage. This can be remedied. The new government of India has embarked on a great plan of agricultural development which will go hand in hand with the complementary industrial program. If the industrial program could be carried to completion in time, the food supply of India could be doubled in 10 or 15 years.

In countries with the necessary technical knowledge and industrial equipment, food output can increase much faster than in the underdeveloped countries. During the war, when there was a market for all that could be produced, the U.S. increased its agricultural production by 35 per cent with 25 per cent less labor, despite soil erosion. In the United Kingdom food production increased by 30 per cent during the war, and the present program aims at a 50 per cent increase by 1952. Better conservation of pastures by grass-drying and silage, is relieving the island of its dependence upon imported livestock-feed, purchased abroad at the rate of eight million tons per year in prewar days. In the past, except during wars, the food production of industrialized countries has been limited not by the capacity of the land but by the buying power of the world market.

But what of the Malthusian axiom that population must increase in geometrical progression? If that held true, the food problem, of course, would be ultimately insoluble; there would not be standing room on the surface of the earth. History shows no such long-term process. The population of Europe, after its explosive growth of the 19th century, entered a new phase in this century. With the rise in the standard of living and education, the birth rate fell to such an extent that some governments, alarmed at the resulting decline in population, offered grants to induce women to have children. The probability is that the same course of events will ensue in the poverty-stricken countries where populations are now increasing at an accelerating rate. In the long run the rise in the standard of living and education all over the world should be followed by a decline in the birth rate to the replacement level corresponding to the reduced death rate.

So much for the conflicting views on whether old mother earth can produce sufficient food for her children. The view of the author is that if she fails and the gloomy predictions of famine by Aldous Huxley and others are fulfilled, the disaster will not be due to her niggardliness. It will be due rather to the failure of governments to adjust international politics and economics to permit her to give of her almost inexhaustible potential abundance.

Let us examine first the political threat of hunger, since it is the most urgent. The estimate that two thirds of mankind suffers chronic malnutrition dates from before World War II. The acute postwar food shortage has been relieved by the increased production of grains, rice, sugar and potatoes from the prewar level of 502 million to 547 million tons (in terms of wheat equivalent). These gains are offset somewhat by the failure of the production of high-protein animal products to recover to prewar levels. But meanwhile the world population has increased by about 200 million. Hence though the production of carbohydrate-rich food has increased, the supply per person is still about three per cent below prewar. Correspondingly the supply of animal products is nearly 10 per cent below the inadequate prewar level. As a result the aggregate calorie supply is about seven per cent less, or about 2,200 against 2,390.

The shortage is the more severe in many regions because the distribution of available supplies is uneven. In the U.S., where before the war the average intake of the more expensive protein foods was already high, the consumption of these foods has now increased 20 to 30 per cent. The reverse is true for most of the food-deficit countries, where both total calorie and protein intake have declined. Indeed, less protein has been available to more than 80 per cent of the world's population since the war. The deterioration of supply has been most severely felt in Asia, the region of greatest prewar need. In this period daily calorie intake per person declined in India from 1,968 to 1,621, in Burma from 2,080 to 1,937 and in Japan from 2,175 to 1,834. The calorie requirement for these warm countries is estimated at about 2,600 per head. In the U.S. the present average consumption is about 3,200.

It was hunger following the bad harvests of 1788 that led to the excesses of the French Revolution of 1789. In the European revolutionary movement of the 1840s—the 'hungry forties'—the mobs in the industrial towns of England chanted 'bread or blood'. Today the hunger and poverty of two thirds of the world's people, who comprehend that their lot can be improved, is a more potent cause of the spread of Communism than fifth-column movements in the U.S. and Western Europe.

Yet in the midst of this dire need there remains the economic threat of the food surpluses generated by the progressive technologies of a few fortunate nations. The abundant food output of the U.S. already imposes a heavy burden upon its economy and now has begun to undermine its prosperity. Up to January 31, 1950, the U.S. Treasury had paid out $2,470 million for taking surplus food off the market. In inevitable accord with this policy, production is now being restricted. The 1950 wheat acreage has been reduced nearly 10 million acres below 1949. A further 10-million-acre reduction planned for 1951 would bring total U.S. wheat acreage down to a little more than 60 million. Since U.S. farmers in previous prosperous years bought industrial goods to the value of $9,000 million per year, reduction in their purchasing power will inevitably lead to urban unemployment. This in turn leads to a reduction in the market for more expensive foods. The consumption of these has, in fact, already declined by five to 10 per cent from the postwar peak, increasing the unmarketable surpluses.

The only way to avoid the catastrophe inherent in this vicious cycle is to adjust international finance and economics to permit the distribution and consumption of the abundance that modern technology can create. The best starting point for action toward this grand objective would be a world food policy based upon human needs. Under such a policy, directed at the abolition of famine and chronic hunger, the world would have no unmarketable surpluses for many years to come.

Taking account of the anticipated increase in the world population, it is estimated that food production would have to be doubled in the next 25 years in order to provide enough for all. Such a goal presents no insuperable difficulties. The doubling of world food production, with a guaranteed market for all that could be produced, would bring prosperity and stability to agriculture all over the world. It would demand vast quantities of industrial products for water shortage and irrigation schemes, for reclamation of land destroyed by soil erosion, for fertilizers and equipment, for means of transport and for consumer goods to supply the increased purchasing power of the vast market of food producers. It would thus bring about business prosperity with full employment in an expanding world economy.

A world food plan on these lines would of course call for the creation of enormous credits. There is ample precedent for international financing on this scale. In the 19th century Britain granted credits and made loans to countries in every continent, including the U.S., with no political strings attached. These credits were repaid through the creation of new wealth. Without doubt another such investment would again be repaid in the future.

It was to this end that the United Nations' specialized agencies were organized. The Food and Agriculture Organization, the Economic and Social Council and the World Bank for Reconstruction and Development were set up to enable nations to cooperate in the development of the resources of the earth for the benefit of all mankind.

But we must contend with the immediate crisis created by the existence of hunger side by side with food surpluses. In 1946 the author, then Director General of FAO, submitted to the governments of the world a proposal for a World Food Board. This agency was designed to provide the financial and other technical arrangements necessary to convert human need into effective demand in the markets of the world. It was to have the power and the funds for positive action.

If it were at work today, the World Food Board would be engaged in operations similar to those conducted by the agricultural commodity agencies of the U.S. Thus it would be responsible for maintenance of stable world prices and would be empowered and financed to buy and sell food on the world markets. Its determination of price levels, however, would involve international rather than national or sectional interests. As a result of its investigations, the Board would calculate prices to call forth the production of food in volume sufficient to meet world demand. Since it would be buying at lower and selling at higher prices, successful management would make it possible to conduct these operations on a revolving fund of reasonable size.

Moreover, the Board would now as a principal function be finding use for the food surpluses of the U.S. and the few other nations that are burdened with them. With the collaboration of the appropriate UN agencies, it would arrange for the financing necessary to dispose of these surpluses on special terms to the peoples whose need for them is most urgent. In addition the Board would have famine reserves on hand adequate for any emergency that might arise through failure of crops in any part of the world. Finally such an agency would inevitably be playing a major role in advancing international effort to double the world's food production.

At the first FAO conference the U.S. approved of the World Food Board proposal, and its delegation moved to set up the preparatory commission to get it going. But when the commission met two months later the U.S. had changed its mind. The World Food Board proposal continued to hold the support of the majority of national delegations, particularly those of France and India, which did everything possible to get it carried through. The best that could be done, however, was to establish a council which has neither the funds nor the authority to get action taken.

Today the U.S., by Marshall Plan aid to Europe and the offer of technical assistance and financial aid to other countries, is trying to do by itself what the UN agencies were created to do. It is doubtful whether it will succeed. To make it work, the U.S. Government would have to expand its program to the $50,000 million scale projected by Senator Brien McMahon. This would impose an intolerable burden upon the U.S. taxpayer. The world cannot be made richer by making the U.S. poorer. Further, the unparalleled generosity of the U.S. can be, and is already being, misrepresented as an attempt to impose American economic imperialism on the world.

To be realistic in advancing the proposal for a World Food Board again in 1950, one must meet a serious political question. Would the U.S.S.R. cooperate? In 1946 the late Fiorello La Guardia, with whom the author was working in the closest contact, reported that he had discussed the proposal with Joseph Stalin and other responsible Soviet leaders. He quoted them as saying that it was the one UN project in which they would participate, once they were convinced that the U.S. and Britain would do likewise and that it would not be dominated in the interests of the U.S. Relations have since worsened, and it is now very doubtful whether the Russians would join. But they should be given a chance. If they refused, the selfish objectives of their national policy would be exposed to world opinion.

Even without the U.S.S.R., the cooperation of the other nations would go far toward removing international trade and financial restrictions. By consolidating the economic position of these nations, such action might prove more important in a continuing cold war than an advantage in the arms race. Further, it would fire the imagination of the peoples presently in revolt and create a new world spirit of hope and a willingness to work together.

The two evils of the glut of food in the U.S. and hunger in widespread areas in the rest of the world cannot be eliminated by U.S. charity. The food problem is international. It can be met only by an international effort. Conducted through the technical agencies of the UN, the world food program would be placed upon a business footing and bring as much benefit to the U.S. as to other countries. Only by multilateral action can the UN redeem the forgotten pledge of the Atlantic Charter to achieve 'freedom from want among the people of all lands.'

7

Public Health and Medicine

7.1   John Simon, *English Sanitary Institutions*, 1890

With regard to the progress of English sanitary administration during the seventeen years between the establishment of the Local Government Board and the passing of the Act of 1888 for the constitution of County Councils, some admixture of disappointment may be confessed; for if, during those years, the Local Government Board had exercised more influence of real supervision in favour of progress, and against the many local inactivities and defaults which have been known to exist,—especially if in the first instance it had been started with organisation and spirit for that branch of work,—presumably the present sanitary position of England would have been much more uniformly advanced than it is.

That drawback, however, has not had power to reverse the generally advancing tide; and, in spite of it, the total of our past half-century's British progress in sanitary knowledge and government has been such as will make the Victorian period memorable for future history. The Science of Preventive Medicine has immensely advanced; not only in the immediate gain, that various diseases, and their respective causes and respective modes of propagation, are far more exactly understood than before, and that the diseases can therefore of course be more readily prevented; but also, and even more largely, because new methods of pathological research have been created, full of the utmost promise for future augmentations of exact knowledge.[1] Popular acceptance of the scientific teaching, and popular experience in confirmation of it, have also made very considerable progress,—enough to have inclined local populations to tolerate with but little grumbling, or even in many cases to promote with more or less zeal, large financial expenditure for purposes of Public Health: the Medical Profession has come to be recognised as an ally of indispensable helpfulness for the State in affairs of both local and general government: while, further, in consequence of the more specific demands which Preventive Medicine has made for various mechanical and physical and chemical aids and appliances, new branches of commerce, purporting to fulfil various dictates of Preventive Medicine, have begun to arise, and in some cases have grown to excellence.[2] On the new foundations of Science, a new political superstructure has taken form. For the purpose of locally protecting the Public Health, a great body of new law, and a vast apparatus of administrative machinery, wherein medical officers form an essential part, have come into existence; the Public has begun to feel its own incalculable

Source: John Simon, *English Sanitary Institutions Reviewed in their Course of Development, and in some of their Political and Social Relations*, London: Cassell, 1890, pp. 462–469, 475–476.

interest,that this new branch of our national politics should be worked with intelligence and honesty; and, more and more throughout England, men, possessed of the qualifications to be desired, have been girding themselves in answer to that appeal. Evidence enough is already to hand, that, where local government has been reasonably attentive to modern sanitary rules, great improvements in local salubrity, great diminutions in the quantity of local disease and death, have, as had been predicted, come to pass.[3]

It is peculiarly gratifying to note that the English progress of the last half-century has been of influence far beyond the limits of the United Kingdom; not only in the colonies and other transmarine parts of the British Empire, but also in countries under other government; and the early English workers, who have joyfully witnessed that wide extension of a great beneficence, have at the same time had the happiness of finding their own pioneer-work approved and honoured by the chief foreign promoters of the extension.[4]

The progress which has been made consists essentially in practical applications of Pathological Science; and happily that branch of knowledge shews every sign of continuing to give lessons for application. In the eyes of those who cultivate it in a spirit of becoming modesty towards the magnitude and the difficulty of their subject-matter, it, no doubt, like many other branches of the infinite study of Nature, appears hitherto as only in that first stage of true growth where the known is immeasurably less than the unknown; but even in this early stage it has already given ample light for very large preventions of disease; and, so far onward as we can foresee, we may expect that its light will continue to be an ever-advancing guide for advances of law and conduct. It is now proceeding with such activity as the world has never before witnessed, and the various kinds of knowledge which supply resources for the prevention of disease are increasing with immense rapidity. Clearly we have to hope that, in proportion as exact knowledge is gained of agencies prejudicial to the public health, the nation will provide against them by appropriate law and by effective administration; but, for obvious reasons, it is not likely that practical reforms will keep themselves immediately abreast of scientific progress. For them, namely, the rate of advance must in chief part depend on the progress of popular education as to the facts and interests and duties of the case, and can therefore hardly be expected to be other than gradual and somewhat slow. Thus it has been that down to the present time, our disease-preventive provisions of law have certainly not in all respects kept pace with what we know as to the causes of disease; and even less advanced in most instances is the readiness of persons and authorities to make full use of the provisions which exist.

Even as regards those parts of the case where popular education might now be supposed to have become comparatively ripe—the parts which specially regard the Cultivation of Cleanliness, it would be flattery to pretend that average England has yet reached any high standard of sensibility to dirt.[5] Against accumulated obvious masses of filth, against extreme ferocities of stench, local protests no doubt are pretty commonly to be heard, and, at moments when there is panic about disease, may often rise to considerable warmth of indignation; but in regard of the less riotous forms of uncleanliness, far too much insensibility is widely shown. See, for instance, how little fastidiousness prevails in the popular mind as to the domestic and commercial arrangements which supply *Drinking Water*. To say nothing of domestic neglects as to the cleanliness of cisternage, and nothing as to the frequency with which private suplies of water are derived from surface-wells sunk in foul soil, and imbibing from cesspools and muck-heaps, merely let note be here taken of the general inattentiveness to questions of *public* supply. Now and then, no doubt, may be heard an expression of mild surprise, that water which purports to be 'filtered' under Act of Parliament has much of the aspect of third-class ditch-water, or is found to have plugged its service-pipe with some live or dead body of an eel; but, of proper watchful insistence that the drinking-supplies shall be systematically guarded from pollution, there is hardly a trace to be found; and the consequences of this carelessness (quite apart here from any question of its bearing on health) are often disclosed in forms of such filth as ought to be blushed for. That even the London water-supply, after half-a-century of disgusting disclosures, and after various very terrible disasters, is not yet secured against gross defilement, is a fact to be sufficiently gathered from the reports of the official examiner under the Metropolis Water Act, 1871, and is in other ways deplorably notorious.[6] In the summer of 1886, the *Lancet* medical journal brought to light that, during the week of the Henley Regatta, the Thames, for about a mile's length of its course, where supposed to be sacred to the water-supply of London, had had, on and about its surface, a floating and riparian encampment of some thousands of holiday-makers, using the river as their latrine and middenstead, and with their house-boats purposely closet-piped into it: all this apparently not anything new, but a story which would perhaps strike the popular mind when the medical journal had commented on it![7] What sentiment of cleanliness prevailed among the thousands who could thus deal with their neighbours' drinking-water, and among the millions who were placidly bearing the outrage, is a question which may be left for such future historians as will discuss the curiosities of English civilisation at the close of the nineteenth century; and in the meantime national education will perhaps have taught that a river, having manured fields and sewage-farms and populous urban districts along its banks, and constituted by law 'a navigable highway on which all persons have right to pass and re-pass for pleasure or profit,' is not (even apart from regattas) likely to supply such drinking-water as moderate sentiments of cleanliness would seem to demand. Again, to look in another direction, see with what apparent indifference our nineteenth-century England acquiesces in a daily-increasing sacrifice of daylight to dirt. There are immense masses of our population—the inhabitants, for instance, of London and of many chief manufacturing towns, who endure without revolt or struggle the extremities of general *Smoke Nuisance*; not only condoning the fact (on which here the argument does not turn) that the nuisance is of painful injury to an appreciable proportion of persons, and in certain states of weather kills many of them; but further (which is here the point) accepting, as if in obedience to some natural law, that their common life shall in great part be excluded from the pure light of day—that incomparable source of all physical gladness—by an ignoble pall of unconsumed soot; and hardly murmuring, in their self-imposed eclipse, that their persons and clothing and domestic furniture are under the incessant grime of a nuisance which is essentially removable.—It is of purpose that I have adverted, though but in few words, to some comparatively non-sanitary aspects of the question of cleanliness: for disease-preventive degrees of cleanliness will hardly be attained, unless something more than disease-prevention be included in the popular aim. A great people, determining what it will deem to be proper purity for air and water, has not to measure only from the scavenger's point of view, but surely also with some sense of the help which accrues to the human mind from beholding the pure aspects of nature, and with some readiness for displeasure when the beauty and bounty of nature are wantonly affronted by slovenliness and waste. For rich and poor alike, it cannot be too clearly understood that the claims of cleanliness are fastidious. In order to sanitary self-protection by its means, there must be sufficient refinement of taste to abhor even minor degrees of dirt, and to insist throughout on the utmost possible purity of air and water; there also must be sufficient sharpness and cultivation of sight and smell, to immediately discover even minute infractions of the sanitary rule; and there must be sufficient intelligence and watchfulness as to the channels, commercial and other, which can clandestinely admit uncleanliness from without.

While average English life is but imperfectly educated in standards of cleanliness, and in knowledge of the dangers which uncleanliness involves, it may be taken for granted that popular education is even less advanced in the other rudiments of sanitary knowledge; and with so widely defective an appreciation of the causes of disease, there of course will be corresponding voids in the practice of sanitary self-protection. The uneducatedness has, for one of its consequences, that, in various ways of neglect and passivity, sometimes by individuals omitting to do for their own persons or premises something which they personally ought to do and sometimes by their omitting to invoke from appointed authorities or offices such assistance or protection as the law intends these to afford, very many incur disease, who, with better education, would know how to escape it, and could in general be expected to exert themselves to that effect.[8] Thus far, however, is mentioned only one of the chief preventable evils against which advancing education has to contend; and there remains for mention a second which is at least equally important. Diseases do not spread only through the *passiveness* of those who suffer them, but spread, in immense quantity, through the influence, essentially *voluntary*, of wrongful acts, neglect, or defaults, on the part of others. It is to be hoped that, with advancing popular education, those aggressive activities of disease-production in various parts of our social system will be seen in stronger and stronger light, and be more and more plainly understood, as very serious forms of social wrong; and that there may accordingly be developed against them, on the one hand, such public opinion as will insist on higher standards of social duty, and, on the other hand, such sufficient stringency and punitiveness of the law as may be a terror to evil-doers. In illustration of my meaning (but of course without intending that the argument should limit itself to the instances cited) I would refer to the frequently-recurring cases in which wrongful conduct creates an Epidemic of Infectious Disease, or widely spreads such disease beyond its first limits; to the cases, where either local authorities or commercial companies, purveying public water-supply, have made epidemics of cholera or of enteric fever, sometimes on a frightfully large scale, by distributing polluted water; to the cases, where local authorities, by non-construction or mal-construction of sewers, or by omitting to make proper application of the nuisances-removal law in their districts, have occasioned similar epidemics; to the cases where dairies or dairy-farms have supplied to masses of customers a polluted milk, infectious of enteric fever, or a milk otherwise infectious; and, not least, to the innumerable instances in which improper conduct on the part of an infectiously-diseased person, or of persons in charge of him, has propagated infectious disease, necessarily with more or less possibility of further personal propagation, and perhaps, sooner or later, of wide dissemination through some commercial or other apparatus having large contact with the public—some school or laundry, or food-shop, or water-source, or dressmaking or upholstering industry, and this improper conduct representing either an almost savage ignorance as to the nature of the disease, or else an utterly selfish indifference as to the hurt one's conduct may cause to others. So, again, to add illustration from another branch of health-government, I may refer to the quantities of disease and death brought upon the public through the almost unbounded facility which exists for abuses and dishonesties in the house-trade, and by the frequency with which jerry-built and other unfit houses, having in them latent malconstructions dangerous to health, are let for hire to persons who have not knowledge enough to protect themselves against the harm. [. . .]

The question of popular education in such sanitary matters as have been under review ought hardly to be left without a word of reminder as to the boundary-line between '*public*' and '*private*' in affairs of Health. While modern times are recognising on a large scale, as general principles, that every community has an interest in the health and strength of its individual members, and that in various important respects the aggregated individuals cannot secure health for themselves unless they act solidly together by appropriate defences of law and administration, those principles do not at all imply that the community relieves its individual members from the general responsibility of caring for themselves, or undertakes to prevent individual acts of unwisdom by which the individual causes injury to his own health. Long before our modern codes of public sanitary law had begun to shape themselves, elaborate counsels of personal hygiene had become current in the world; counsels, as to the ways and habits of life which would most conduce to healthful longevity; counsels, above all, for moderation in life—'the rule of *not too much*;' and those counsels for personal self-government, enforced from age to age by the ever-growing common experience of mankind, are not now to be deemed superfluous because boards of local government have arisen. In relation to the sexes and their union, and to the many personal influences which are hereditary, in relation to eating and drinking, in relation to work and repose and recreation for mind and body, in relation to the charge of infancy, and to proper differences of regimen for the different after-periods of life, there are hygienic rules, perhaps not less important to mankind than the rules which constitute local authorities; but to enter even slightly on the discussion of them would be far beyond the intention of these pages; and I cannot in passing do more than thus advert to the importance of the subject.

Notes

1   Observe, for instance, even since twenty years ago, how the knowledge of the infective diseases has been increased by cultivation of the morbid contagia in artificial media, and by more discriminative methods of experimental inoculation, and by exact comparative studies of the powers of alleged disinfectants.

2   Compare, for instance, the details of present house-drainage, as laid down under competent direction, with the details which passed muster in times before the General Board of Health. Or compare the present time with thirty years ago in respect of the means (commercial as well as municipal) which can be invoked, to assist in preventing the spread of infectious diseases.

3   Details of such evidence are to be found in very many reports of local Officers of Health. For more collective statements, see Dr Ogle's highly instructive *Letter to the Registrar-General*, in decennial supplement to the latter's 45th Annual Report. See also, in the 50th *Annual Report* of the same department, the Registrar-General's reflexions on the past *Fifty Years of Civil Registration*; and, in the just-published 51st *Annual Report*, the encouraging statements which are made as to the death-rates in this current decennium.

4   It would be most agreeable to me were it in my power to offer here some analysis of the sanitary literature of the last twenty or thirty years; to show how much merit there has been in many admirable reports which have been issued under local administration in the United Kingdom, and under our colonial and Indian jurisdictions; and at the same time also to tell something of the activity which has been shown in other countries—as, above all, in Germany, and in the more advanced of the United States of America; but with my present limits of time and personal strength, I dare not enter upon any so ambitious attempt. If I may permit myself partial exception where necessarily my chief rule must be silence, there are two foreign names which all contemporary opinion will, I feel sure, justify me in mentioning with peculiar respect: the names of two, whose life-long devotion to the advancement of sanitary science and practice, has laid all us their fellow-workers under obligation; one, the happily still living and vigorous Munich Professor, Dr Max v. Pettenkofer; the other, Dr Georg Varentrapp, the lately deceased patriot citizen of Frankfort. I may add that, during the time, the chief English Medical Journals have in general guided their readers to what has been of most interest in current British work; and that all chief sanitary thought of the last twenty years has been well represented in the valuable *Deutsche Vierteljahrschrift für öffentliche Gesundheitspflege*, now edited by Dr Alexander Spiess and Dr M. Pistor.

5   When I speak of *average* England still having to learn lessons, and even rudimentary lessons, in various matters of sanitary cleanliness, I do not intend to imply that the wealthier classes of society are an exception to that average reproach. It is by no means alone in comparatively poor and ungarnished dwellings, that filth-diseases and odours of filth are to be found. In the houses of wealthy and self-indulgent persons, who perhaps may be spending money and raptures on the fine arts, and who certainly would think it strange to find themselves under imputation of dirt, and in the highly-paid lodging-houses which these classes inhabit from time to time at their so-called health-resorts, it is not very rare—indeed, as to the lodgings, it is rather frequent, that the staircase is pervaded by more or less sewage-odour from defective drain-structures thereabouts or in the basement; and even the wealthiest know but too well that enteric fever, with its congeners, does not leave them unscathed. Persons, not fairly educated to profit by their sense of smell, stumble as naturally into certain sorts of disease as the more or less blind stumble into other pitfalls; and a suggestion that the non-observant *Johnny Head-in-Air* will come to grief in matters of

scent, as well as in matters of sight, is one which some ingenious future *Struwelpeter* might fitly endeavour to bring into his nursery picture-book.

6   See, for instance, with particular reference to certain intakes from the Thames, the Report on September, 1888.

7   See *Lancet*, July 17, 1886; and, for such amendments as followed on that exposure, see the reports of the same journal in succeeding years.

8   *In general*, I say: for (as at least one necessary qualification) it must always particularly be remembered, that persons in various forms of social dependence may be in extreme difficulty as to their power of claiming sanitary rights, and obtaining emancipation from causes of disease.

7.2   H. Florey, *Penicillin*, 1944

The chemotherapeutic properties of penicillin were discovered in 1940, but before this there was a long history which for convenience can be divided into stages. (1) The discovery of naturally occurring antibacterial substances—or antibiotics, as they are now beginning to be called—and the early attempts to utilize them in medicine. (2) The discovery of the antibacterial substance penicillin by Alexander Fleming. (3) The discovery of its chemotherapeutic properties at Oxford. (4) The stage of development in which we are at present, which consists of three interrelated lines of research—namely, (i) exploration of methods for mass-producing penicillin by the growth of the mould *Penicillium notatum*; (ii) investigation of the chemical structure of penicillin with the hope that it may eventually be synthesized by chemical means; and (iii) the clinical exploitation of the known properties of penicillin.

Stage 1

We have to go back to 1877 for the first observation of a naturally produced antibacterial substance. In that year Pasteur and Joubert described how when common air bacteria contaminated flasks of broth containing the bacillus of anthrax the growth of the anthrax bacillus was stopped. That phenomenon was probably the first observation that one organism may produce a chemical substance—or antibiotic—which is capable of stopping the growth of another, though Pasteur did not realize its true significance. In the succeeding years many examples were discovered, of which the most interesting was *Bacillus pyocyaneus*. From the medium on which this organism had grown Emmerich and Loew extracted a substance which they called pyocyanase. This was found to be capable of stopping the growth of certain organisms causing disease, notably anthrax and diphtheria. They applied it to the lesions of the skin caused by anthrax with, they claimed, some benefit. Although this product was on sale in Germany as recently as the 1930's its use in medicine never became widespread.

Source: H. Florey, 'Penicillin: a survey', *British Medical Journal*, 1944, II, pp. 169–171.

Stage 2

In 1928 Fleming was studying the staphylococcus. One day he examined and then put aside on his bench a plate on which colonies of the staphylococcus were growing. Several days later there was a colony of mould growing on one side. Fleming noticed that in the neighbourhood of the mould the colonies of staphylococci were disappearing. He recognized this as a phenomenon of interest, and subcultured the mould, which was later identified as *Penicillium notatum*. When grown on nutrient broth it was found to produce some substance which passed into the liquid. By experiments in test-tubes Fleming showed that the liquid had the property of stopping the growth of many bacteria. Fleming called the active liquid penicillin. He carried out experiments on the effect of his broth on numerous organisms in test-tubes and showed that many which can cause disease in man were affected, although some disease-producing organisms were quite insensitive. He also injected some of the broth containing penicillin into rabbits, and found that it was no more toxic than ordinary broth. He found, too, that the broth did not harm the white blood cells. Fleming, who had been working on antiseptics, recognized that penicillin had some very desirable properties as an antiseptic, and proposed that it might be useful for local application to infected surfaces. He did, in fact, so apply it in a few cases, with results indicating, as he said, that 'it certainly appeared to be superior to dressings containing potent chemicals.' About this time an attempt was made by Clutterbuck, Lovell, and Raistrick to extract the penicillin. They succeeded in growing the mould on a purely synthetic medium and found that the active substance could be extracted into ether when the watery medium containing penicillin was acidified. However, when they tried to concentrate the penicillin by evaporating the ether most of the activity was lost, and they concluded that penicillin was 'extremely labile.'

We may briefly summarize the position at the end of this phase by saying that Fleming had discovered the existence of an antibiotic produced by *Penicillium notatum*. Some test-tube investigations had been made of the antibacterial power of the crude broth and, as a result, it had been suggested that it might be useful as an antiseptic locally applied to infected lesions. But as the result of both Fleming's and Clutterbuck, Lovell, and Raistrick's work the conclusion had been reached that penicillin was an unstable substance and therefore unlikely to have any practical value in medicine.

Stage 3

Stage 3 deals with the work done at Oxford. My own interest in the phenomena of bacterial inhibition began in the 1920's. Since 1929, at first alone and later with collaborators, work had been in progress, but it was not till 1938 that Dr Chain, a biochemist, and I prepared a plan for the systematic study of some of the naturally produced antibacterial substances. After much discussion the choice was narrowed down to three—*Bacillus pyocyaneus*, *Penicillium notatum*, and the subtilis-mesentericus group of bacteria. Eventually work was undertaken on the first two. Miss Schoental obtained three antibacterial products from *Bacillus pyocyaneus*, which all proved to be very toxic, but fortunately the results with penicillin turned out rather differently.

Both Fleming and Clutterbuck, Lovell, and Raistrick had noticed that under certain conditions the crude broth might retain its activity for at least several weeks. This indicated that in appropriate conditions the substance might not be so unstable as had been pictured. To work on the metabolic products of moulds from biological as well as chemical aspects needs a team of specialized workers, so that the various fields of investigation may be covered, and it was most fortunate that such a team was available in Oxford at that time. I should like to stress that this work could not have been carried through had it not been for the unremitting labours of the following people: Dr Chain, Dr Abraham, Prof Gardner, Dr Heatley, Dr Jennings, Dr Sanders, Dr Fletcher, and Lady Florey, Nor could we have got far without the work of our technical assistants. Mr Glister and his 'penicillin girls,' Mr Kent, and, for the chemical work, Mr Callow and Mr Burtt.

The body of work done by this team in the next two years produced a single end-result—penicillin as a proved chemotherapeutic drug. The steps cannot be set out chronologically because different aspects of the work were in progress simultaneously, and the accent was first on one thing and then on another until a fairly complete picture was built up. The first step in all work of this type is to grow the mould on a medium into which it will produce the active substance. Here we were able, in the first place, to use the information which had been obtained by Fleming and by Clutterbuck, Lovell, and Raistrick, and we began by growing the mould on the synthetic medium proposed by the latter workers.

In studying an antibiotic its fundamental property of inhibiting bacteria can be made use of as a test method, and we owe to Dr Heatley the elaboration of a test which has proved invaluable for work not only on penicillin but on many other antibiotics as well. By means of his method it was possible to follow the various fractionating processes. The crucial chemical observation was the demonstration that not only did penicillin, when made acid, pass from a watery into an organic solvent such as ether or amyl acetate, but that it could be recovered from the organic solvent when shaken with water and an appropriate amount of alkali. By repetitions of this process purification and concentration were effected and the first stable products containing penicillin produced.
[. . .]

The fact that penicillin is a very powerful antibacterial agent would not by itself differentiate it from a number of other mould products or from some of the familiar chemical antiseptics. But whereas nearly all such substances are quite toxic to body tissues, even concentrated extracts of penicillin had practically no poisonous action on animals. It was further shown that individual body cells, such as the white cells of the blood were unaffected by concentrations many hundreds of times greater than those necessary to stop the growth of sensitive organisms.

When administered to an infected animal or man in sufficient quantity penicillin stops the growth of the germs, thus giving the white blood cells in particular, and possibly other defence mechanisms, the opportunity effectively to attack and destroy them. It was found, too, that tissue cultures would survive and grow in concentrations very much greater than those necessary to produce bacteriostasis. In animals the active material was rapidly excreted by the kidneys into the urine, and, to a lesser extent, by the liver into the bile. It was readily absorbed after injection under the skin or into the muscles or into the small intestine, but it could not be given by mouth because of the hydrochloric acid in the stomach, which destroys penicillin very rapidly. Neither could it be given by the large bowel because the bacteria there destroy it.

The position at which we had now arrived was that we had in our hands a substance which combined very low toxicity to animals with a very powerful action against disease-producing bacteria. We knew a good deal about its fundamental behaviour in the animal body. The most important step had now been reached—we had still to learn whether it would cure disease in animals and man.

It is worth while to digress for a moment in order to take up the question of antiseptics, so that the real significance of the experiments about to be described may be understood. Everyone is familiar with antiseptics such as mercuric chloride, acriflavine, dettol, lysol, etc. All these are capable, under appropriate conditions, of killing bacteria—mark the word 'killing'—but cannot be used for injecting into the animal body because they have a damaging effect on animal cells as well as on bacteria. All the antiseptics in common use destroy protoplasm quite quickly, and this applies equally to the protoplasm of the bacterium and of the animal. As might be expected, although antiseptics can be used for sterilizing instruments and similar purposes, little success has attended their use in dealing with infected wounds, still less their injection into the body. A chemotherapeutic agent differs from antiseptics in that it selectively attacks the organisms causing the disease, without at the same time doing any serious injury in the body. For this reason it can be given internally or by injection. There are several examples of such chemotherapeutic substances. The one which has been known the longest, and is perhaps the most familiar, is quinine, used to combat malaria. Quinine is swallowed by mouth, passes into the blood stream, and exerts its beneficent action in killing the malaria parasite while being carried round to all parts of the body. Another example is salvarsan, the discovery of Ehrlich, produced after many years' work. It is an arsenic compound which has a very profound effect on the spirochaete of syphilis without being too toxic to be borne by the person suffering from the disease. Other substances were discovered which were effective against various tropical diseases, but only one class of substance, the sulphonamides, had been found of any use in common diseases such as sepsis. Their use was, for various reasons—some of which have been mentioned earlier—somewhat limited. These are all true chemotherapeutic agents, not antiseptics.

The following experiments demonstrated that penicillin belongs to the class of true chemotherapeutic agents. So far as the use of penicillin in medicine is concerned this was the crucial discovery. Such experiments are carried out in the following way. Mice are injected with bacteria such as streptococci and staphylococci so that they will certainly die from the infection within one or at most two days. To show that a substance suspected of having chemotherapeutic properties is active it is necessary to secure survival of a substantial number of mice which would otherwise certainly die.

In the case of penicillin this was accomplished by injecting some penicillin under the skins of the infected animals every three hours for several days. The drug was absorbed from beneath the skin into the blood stream, which carried it to the place where the infecting bacteria had previously been placed. Knowing that penicillin was a soluble substance quickly distributed round the body, that it was not toxic to animal tissues, and that it was just as active in the presence of body tissues as in a test-tube, we were justified in hoping that it would stop the bacteria growing as effectively in the body as it did outside. And this proved to be the case. The groups of treated mice survived almost without exception, while the untreated mice all died. These first experiments indicated without any doubt that penicillin belonged to that rare class of drugs which can be used as chemotherapeutic agents.

From this demonstration it appeared that penicillin was likely to have very great potentialities in the field of human medicine. Penicillin at that time was extremely difficult to produce in substantial quantities, so that some time passed before we were able to show its powers on man. We again have to thank Dr Heatley and his assistants for unremitting work in producing in the laboratory enough penicillin for the first injections in man. Even after months of work we could treat only six cases of severe infection, but the results were most promising.

The first human patients were treated in the winter and spring of 1940–1, at the time of the worst bombing of England. It seemed improbable that much headway could be made in getting large-scale production started in this country. In these circumstances Dr Heatley and I went to America, which was not then at war, to ask them whether they could put some of their great resources into the production of penicillin, so that more extensive clinical trials could be carried out. We were extremely fortunate in coming into contact with Dr Coghill, Director of the Fermentation Division of the Department of Agriculture's excellent research laboratory at Peoria, in Illinois. The work which he and his colleagues have done on the selection of high-yielding strains of *Penicillium notatum*, and on the modification of culture media, has greatly increased the yield which can be obtained from the mould, and has played an important part in the large-scale production of penicillin.

While this work was being initiated in America, enough material was made in Oxford and by Imperial Chemical Industries to enable some eighteen patients with severe infections, most of them caused by the staphylococcus, to be treated. These results were again of such great promise that any effort to produce the drug on a really large scale was clearly worth while. This was more so since certain of the bacteria susceptable to penicillin cause some of the most common and universal infections, including those of war wounds.

Stage 4

From that time the work branched in three directions. First, it was clear that it would be very desirable to make the substance synthetically by chemical procedures without the intervention of the mould. Work is now proceeding along these lines in Oxford, where Dr Chain and Dr Abraham are collaborating with Sir Robert Robinson and his colleagues, and elsewhere, both in this country and in America, hundreds of chemists are engaged on this important problem. Progress in this direction cannot be reported as it is now in the secret category, but the fact has already been published that pure penicillin has been obtained. This was done in America and in Oxford at about the same time. Every resource has been mobilized to deal with this chemical question, but whether success will attend the effort to produce penicillin by synthesis it is impossible to forecast.

The second, and more immediately practicable line, has been to increase the manufacture by means of the mould to a really large scale. This has involved a large number of intricate technical problems, which have been tackled along different lines by the various commercial firms, both in this country and in America. As a result of their efforts penicillin can now be issued by the kilogramme, although, of course, the supplies still fall lamentably short of the demand.

The third line has been to explore further the use of penicillin as a curative agent. There are two possible ways of using penicillin. First, it can be injected into the muscles or veins so that it is carried around in the circulation to the parts which are being attacked by the infecting bacteria. This method is obligatory in the more serious and widespread diseases such as pneumonia, diseases of the bones, and septicaemia, where the diseased tissues cannot be reached by any other means. Although in many cases this is a very effective method, it has the disadvantage of requiring relatively large amounts of penicillin, since the drug is rapidly excreted by the kidneys. Secondly, penicillin may be used as a local application to the affected part. This can be undertaken only if every portion of the infected tissues can be reached by the penicillin, and a good deal of the success of local application depends upon surgical ingenuity in ensuring that the penicillin, which is rapidly absorbed from a wound, is kept in contact with all the infected tissues long enough to exert its action. At the present time a great deal of thought and study is being given to the problem of war wounds and how best to utilize penicillin, both locally and generally, for their treatment.

The increasing supplies of penicillin now available permit of extensive explorations of its use in many diseases. Perhaps the most striking recent addition to knowledge is that of the Americans, who have discovered that penicillin is apparently effective in treating syphilis. Another excellent development since larger supplies have become available is that penicillin can now be given as a preventive instead of as a last resort. In battle casualties especially, the effort is being made to prevent serious sepsis from developing by giving penicillin at a very early stage.

7.3   René Sand, *The Advance to Social Medicine*, 1948

The characteristics of social medicine
Social medicine, like social policy and human economy, is founded on one

Source: René Sand, *The Advance to Social Medicine*, London: Staples Press, 1952, trans. Rita Bradshaw, pp. 559–565; original French version: *Vers la médicine sociale*, Paris: J. B. Baillière, 1948.

cardinal principle: that no one should at any time lack the means necessary to maintain complete physical and mental health. Social medical action is aimed at ensuring that this optimum is constantly available to all.

Social medicine should pursue a policy of collaboration and coordination between doctors, nurses and social workers, between voluntary organizations, public services and legislative bodies. Jacques Parisot, one of its leading exponents, has set a magnificent example of such co-ordination in Meurthe-et-Moselle. 'The organization of social medicine should not be a matter of theory: certainly, it should be inspired by general principles applicable to all cases, but it should nevertheless be adapted to local conditions and needs, should consider different habits and customs. Furthermore, it should never be content to remain static; it must be vital and free from all rigid convention.'[1] At the same time, as Tuntler points out, it is a part of medicine and should remain under the control of medical men.[2]

Ewald's view of social medicine is strictly national. Needs, their remedies, and the mentality of the people certainly differ from country to country; by this standard, hygiene can also be said to be national. But the general principles, the general trends, the methods, are valid the world over. Like all other sciences, social medicine has been built up of contributions from every part of the globe and this variety is enough to keep it receptive, supple, dynamic, free from the influence of systems, parties, schools of thought; it must retain its objective, experimental, non-doctrinal character. Furthermore, as we have attempted to show in this study and as Sir George Newman has already said, social medicine existed even before history began and its emergence today is simply the result of its evolution from a negative to a positive force.[3]

In the inaugural lecture of his series given at the *Collège de France* Etienne Burnet[4] dealt with the subject of social medicine as the complement of experimental medicine; the latter finds the cause and the remedy, the former must make this remedy available to all.

The concept of social medicine corresponds to that of social disease. There are few diseases which confine their effects to the individual, but there are some which simply cannot be studied or treated without passing beyond the individual to the group, the family, the nation, humanity as a whole. The sick person is always an individual, but sickness is related to society inasmuch as this is regarded as something more than a mere numerical aggregation of individuals—the body social embracing the body physical which is the human organism—the macrocosm as opposed to the microcosm. The diseases which we term social are mainly infectious diseases such as tuberculosis, syphilis and leprosy; but there are others which are not infectious, such as alcoholism and cancer. Their spread, their hold, their development, depend on the solidarity and density of the human group; but also on the fact that society is divided into categories or classes differing widely from one another in their modes of life and therefore in their powers of resistance to disease. Social medicine acknowledges the fact that there are rich and poor and that, if diseases are to be combated, these inequalities must be made good. It has even been said, with some truth, that social medicine consists in effecting 'class-equality in relation to health.' By definition, therefore, it is at once apparent that it is not simply biological but also economic and, as we shall see, psychological. Social diseases retain their hold through psychological, as well as economic, causes.

Economic causes: these diseases persist because, for example, the people are poor; their clothing provides inadequate protection against the cold; their houses are insanitary, dark, overcrowded; their diet is deficient, either in quantity and therefore in caloric value, or in quality, i.e. it is wrongly proportioned between the different types of food, is monotonous, and lacks essential vitamins and mineral salts. It has been proved that these deficiencies lower a people's resistance. . .

Next, the psychological causes. The main one of these is the contradictory attitude that man has always held; he values health but makes only unsustained and half-hearted attempts to preserve it. . . Now medicine can do nothing without the co-operation of the public. Society should educate its members to a *general will* for health of the community which will overrule the individual will.

The suppression of endemic and epidemic disease is much less a question of individual medicine than of social medicine. The doctor may cure his individual patient but he does not uproot the disease, which belongs in the environment, in nature, in society, in our mode of life.

Modern medicine is *preventive*, and preventive medicine's characteristic method is periodic examination of all subjects.

It is also *positive*. Health is an asset which should not merely be restored but cultivated. Medicine can no longer be a mere first-aid service for urgent cases, like a fire-brigade; it must utilize every resource that nature offers for the cultivation, that is the preservation and continuous re-creation of health, our most precious possession.

Modern medicine is *collective*. Knowing that disease will not be suppressed by a growing total of individual cures, it strives to reach the individual through the group and to create such conditions of all-round civilization as will imply the health of individuals. It is collective in both terms of the equation, i.e. the doctors who give the treatment and the public who receive it; doctors and public are both organized into groups.

Medicine's most characteristic feature is the attitude which relates everything to man and studies the environment in which he lives only by relation with him. Man was the main concern even during the phase of main-water supplies, sewerage and disinfection techniques. In the following phase, hygiene became once more human hygiene and turned again to medicine, with its clinics, health visitors, social workers, preventive examinations, maternity and child welfare. Today we are going through the last phase of union between sanitary hygiene and human medicine; problems of nutrition and housing—school meals, rational feeding of industrial workers—are the main preoccupations. There is a tradition of humanism which is the very soul of medicine. [. . .]

The frontiers of social medicine

Where are we to place the boundary line between clinical and social medicine? So long as the doctor remains true to the Hippocratic tradition, and takes no account of his patients' occupation or social condition, he is practising pure clinical medicine; once he begins to worry about the occupational, economic and domestic element, he is using social medicine.

Then again, where does social medicine end and industrial medicine begin? If this latter term is understood in its widest sense, the two become almost synonymous; if industrial medicine is taken to include the housing of the working classes, their diet, their medical services, their non-occupational diseases and even their infant-mortality, and if all social legislation is called industrial legislation, then social medicine is swallowed up in industrial medicine. On the other hand, the contrary point of view, whereby industrial medicine is regarded as a part of social medicine, is the more logical, because occupation is only one of the many social factors by which health is affected.

Industrial or, better, occupational medicine has, however, attained an independent status justified by its peculiar character and the large field which it covers: industrial physiology and psychology, the pathology of industrial accidents and occupational diseases, industrial hygiene, are all recognized branches of medicine. Yet, although many workers' diseases not specifically occupational in character—and they are by far the most numerous—have, in certain cases, some connection with occupation, they mainly arise out of conditions governed by the wages received; and this is a field claimed by social medicine, to be shared when necessary with the other branch.

Other frontier-territories, or what the Germans used to call *Grenzgebiete*, are common to social medicine and other branches of medical study. Is eugenics, described by some as racial biology or racial hygiene, to be included in the framework of social medicine? If it could be established that the hereditary potential diminished from the upper to the middle class and from the middle to the working class, we might legitimately speak of a branch of social eugenics. But, in fact, there is no proof that this is so; whenever poor children have been transferred at an early age to normal conditions, their *average* physical and mental development has proved to be equal to that of children from wealthy homes. It is quite true that among the ranks of the completely destitute there are a great many individuals with bad heredity, but this is only a side-aspect of social medicine.

Moreover eugenics, like industrial medicine, has acquired a certain independence due to the particular character of its objective and methods and the specialized study and experience which it requires.

The same considerations apply to criminology which shares some common ground with sociology, law, anthropology, psychology, psychiatry, and forensic medicine.

Professor J. Leclercq[5] is of the opinion that, temporarily at least, social medicine is bound up with forensic medicine because both work together for the protection of individuals and of society; both are based on legislative measures; both, being deprived of homogenous material proper to their own study, are made up of data and techniques borrowed from various sciences and various other branches of medicine; medical practice comes within the province of both; both are interested in criminology, industrial medicine, social assistance, insurance and social policy.

This, however, is equally true of hygiene, which is even more closely related than forensic medicine to social medicine. The work of medical experts does, nevertheless, make a common meeting-ground for forensic and social medicine.

Professor Fribourg-Blanc, who is attached to Val-de-Grâce, maintains that military medicine is social in essence. It is quite true that it aims even more at preventing disease and developing maximum health and efficiency than it does at providing curative services. This is what Sir Alexander Hood has described as 'total' medicine:[6] as he points out, the health service of the British Army comprised, even before the last war, every medical speciality, including psychiatry, dentistry, even obstetrics and paediatrics, for it covered the soldiers' wives and children, the women of the auxiliary services and women workers employed by the Army. It included school medical services, industrial medical services, prison medical services, medical research services. It dealt with vocational guidance and selection, with vocational training and retraining; of unfit recruits sent to the Physical Development Depot, 87 per cent used to regain normal physical condition in nine weeks of good food, plenty of exercise, plenty of sleep and healthy recreation. The maternal mortality rate among soldiers' wives was much lower than the rate for the general population. Military medicine has been able to obtain these results for the portion of the population under its control; social medicine hopes to extend them to the rest.

Finally, professional medicine and, in particular, that branch which in America is known as medical economics or health economics, is of primary importance to social medicine. Social medicine is not however attached to any particular system; it confines itself to demanding the best preventive and curative services for the whole population. It is not social medicine which has displaced the family doctor in favour of the specialist nor, by making diagnosis and treatment at once more efficient and more expensive, has complicated the problem of medical costs. It is not social medicine which vacillates between regimentation and liberty in the nations' professional and political organization. It was not social medicine which began the 'conflict between a new reality and the old form.'[7] It was not social medicine which transformed living conditions by industrialization, the expansion of town life and the degrading of the craftsman into a member of the proletariat, nor which created the problem of the middle classes. It simply demands that the doctor should understand and perform his duty towards society.

Notes

1   Jacques Parisot, 'La médecine sociale. Ses buts. Principes et méthodes dont elle doit s'inspirer' (*Archives de médecine sociale*. Brussels, June 1938. Vol. I, No. 6, pp 491–511)

2   J. H. Tuntler, *Eadem, sed aliter*, Wolters, Groningen, 1947, pp 8–9.

3   Sir George Newman, 104th Annual Meeting of the British Medical Association (*British Medical Journal*, August 15th, 1936. No. 3945, pp. 358–359).

4   Etienne Burnet, 'Médecine expérimentale et médecine sociale' (*Revue d'hygiène et de médecine préventive*, May 1935, Vol. LVII, No. 5 pp. 321–342).

5   J. Leclercq, 'A propos de la médecine sociale' (*Annales de médecine légale, de criminologie et de police scientifique*, April, 1936, Vol. XVI, No. 4, pp 173–177).

6   Sir Alexander Hood, 'Total Medicine' (*Lancet*, June 9th, 1945, Vol. CCLVIII, No. 6354, No. XXIII of Vol. I. 1945, pp 711–715).

7   Henry E. Sigerist, 'The Medical Student and the Social Problems confronting Medicine Today' (*Bulletin of the History of Medicine*, Baltimore, May, 1936, Vol. IV, No. 5, p. 412).

7.4 Great Britain Ministry of Health, *Report for 1948–49*, 1949

A retrospect

The National Health Insurance Scheme came to an end on 5th July, 1948, as one of the measures repealed by the National Insurance Act, 1946 and the two functions with which it was mainly concerned, the payment of money benefit during periods of incapacity and the provision for free medical attention, became separated in the new social legislation. Responsibility for the administration of sickness benefit passed to the Ministry of National Insurance, whilst medical benefit, now based under the National Health Service Act on an entirely different concept, remained with the Ministry of Health. Though these two aspects are closely related, the separation became inevitable when it was decided that the health services should no longer be based on insurance contributions. In this way, many of the difficulties which had prevented the National Health Insurance Act from achieving its ultimate aim—'the prevention and cure of sickness and for purposes incidental thereto'—should be overcome. The National Health Insurance Act, therefore, marks perhaps one of the most important stages in the history of the general practitioner services of this country, and an appreciation of some of the contributions it has made seems opportune.

Prior to January, 1913, when medical benefit under the Act became operative, a system of contract practice had been widespread, which provided for general practitioner treatment to a considerable proportion of the working-class population, and often to their families as well. The organisations, which contracted with doctors for these general practitioner services, were diverse and included works clubs, friendly societies, medical aid societies and provident dispensaries. The doctor was sometimes employed whole-time and sometimes according to the number of patients he contracted to attend. In the case of works clubs the doctor might be nominated by the employer, or there might be mass meetings of employees for the election of doctors. Some of these arrangements worked smoothly to the benefit of patient and doctor alike, but it was obvious that the system in some areas had developed very undesirable features. A review of the situation was undertaken by the British Medical Association, which published a comprehensive report in 1905. This report brought to light many abuses, the suggestion being made that unfair advantage was taken by these non-medical organisations of competition between doctors. Remuneration was inadequate, and the system frequently reacted unfavourably on the standard of medical care of the sick. A few clubs for instance paid the doctor 1s. 6d. per annum including medicine for each patient, and the amount rarely exceeded 5s. One of the recommendations of the British Medical Association's report was to the effect that practitioners should organise themselves locally to negotiate the conditions under which they undertook contract practice, including rate of remuneration, tenure of

Source: Great Britain Ministry of Health, *Report for 1948–49*, London: HMSO, 1949, pp. 136–140.

office and terms of service. There followed a determined effort by the Contract Practice Committee of the Association to put an end to the defects by proposing a scheme called the Public Medical Service, which was to be organised professionally. The publication of the Poor Law Commission's Reports delayed the implementation of this scheme, and the National Health Insurance proposals prevented any further development.

This preliminary work determined the points on which the profession were determined to stand firm in any new general practitioner service and, during the discussion stages of the National Health Insurance Bill, six 'cardinal points' were advanced: these related to income limits for those entitled to medical benefit, free choice of doctor, administration of medical benefit by insurance committees and not by approved societies, methods of remuneration and representation on committees. These points were all accepted and, apart from some dissatisfaction at the degiee of representation on insurance committees, the only subsequent serious divergence of views with the Government was on the size of the capitation fee. The immediate effect of the National Health Insurance Act was, therefore, to abolish those features of contract practice which were detrimental to doctor and patient alike, and two aspects are worthy of special note. Competition as between practitioners remained, but it depended not on some appointment to a club or other contracting body, but on the doctor's ability to attract patients by virtue of the services he was prepared to give. The free choice of doctor and the standard scale of capitation fee eliminated much of the unhealthy atmosphere of the previous local arrangements in some industrial areas. The idea of the 'family doctor' was retained in that the doctor as a rule included in his practice those families where the insured members were on his list. The second aspect was that of representation on insurance committees and the formation of local medical committees and panel committees, by means of which doctors were able to exert an influence on the local administration of medical benefit and to deal with matters that were considered to be their domestic concern. Insurance committees were constituted for each county and county borough and numbered one hundred and forty-six in England and Wales. The number of members varied between twenty and forty and included two practitioners appointed by the local medical committees and one practitioner appointed by the county or county borough council. Local medical committees were likewise constituted for each county and county borough. Membership of these committees was not confined to insurance practitioners, but was representative of all members of the medical profession in the area. This committee was consulted by the insurance committee on all general questions affecting administration of medical benefit.

Another committee consisting entirely of doctors, of whom not less than three fourths were insurance practitioners, was also formed for each area. This was called the panel committee and it enabled the insurance committee to obtain the views of the insurance practitioners of these areas on certain matters affecting the administration of medical benefit. Questions of discipline and complaint against insurance practitioners were dealt with by a sub-committee of the insurance committee, called the medical service sub-committee. Here there was equal representation of insured persons and the medical profession with an independent chairman. It will be seen therefore that the share accorded to the medical profession in the administration of the scheme was considerable, and in this respect differed much from continental systems of health insurance. There is little doubt that this feature of the scheme accounted in a large measure for the general feeling of satisfaction which was evident amongst insurance practitioners.

The administration of sickness benefit was the responsibility of approved societies, and reference is here made to this part of the subject on account of the control the societies exercised over surplus funds, which were distributed to their members in the form of 'additional benefits'. Approved societies consisted of self-governing associations of insured persons who united to form a society for the purposes of the national health insurance. At the inception of the scheme certain suitable bodies were already in existence, which had been formed by the friendly society movement, and to these were added those instituted by the large insurance companies, those organised by the trade unions and others of less importance numerically. There were over nine hundred approved societies in England and Wales, and of these twenty-eight had branches, each branch having a considerable degree of financial and administrative independence. There were about 6,000 such branches. The number of members of each approved society varied from 50 to over 2,000,000. The health insurance side of the Societies was conducted on a non-profit making basis, and, depending on the amount of sickness claims, surplus funds might accumulate, which were available to members to meet expenses incurred for certain forms of treatment outside the range of the general practitioner. These additional benefits took various forms, the more important of which were payment to hospital for in-patient treatment, payment of part or whole of the cost of dental treatment and payment for the whole or part of the cost of ophthalmic treatment. Help was also given to meet expenses of convalescence, home nursing and surgical appliances. The administration of these additional benefits raised many problems, particularly with regard to dental and ophthalmic treatment. Arrangements made with the dental profession proved satisfactory having regard to the number of eligible members requiring such treatment, though the number of claims for dental benefit was disappointingly small. This was probably due to the fact that approved societies rarely paid for the whole of it. Help was given to the approved societies in the administration of this benefit by the appointment to the Ministry staff of regional dental officers. As regards sight-testing and the supply of glasses the approved societies allowed members to obtain glasses from opticians without producing a medical prescription. In 1923 a Joint council of Qualified Opticians was formed which drew up a register of opticians having approved qualifications. Arrangements were made with many approved societies to send clients to opticians on this register. In 1937, an Ophthalmic Benefit Approved Committee came into being which gave guidance to approved societies and to opticians on questions arising out of ophthalmic benefit and investigated complaints.

Reference should here be made to the Royal Commission on National Health Insurance which was appointed in 1924 with the following terms of reference 'to inquire into the scheme of National Health Insurance. . .and to report what, if any, alteration, extensions or developments should be made in regard to the scope of that scheme and the administrative financial and medical arrangements set up under it'. Comprehensive majority and minority reports were published in 1926, and of special interest in view of subsequent developments of a national health service are the chapters on the development of the health service and proposals for extending medical benefit. The report stated 'we can however, say at once that we are satisfied that the Scheme of National Health Insurance has fully justified itself and has on the whole been successful in operation'. The desirability of greater co-ordination of the health services of the country was frequently stressed, and, as one contribution to the 'unification of local effort on health services', it was recommended that insurance committees should be abolished and their work taken over by the local authority. Of great interest was the following extract from the evidence given by the British Medical Association:

In the year 1922 both the Representative Body of the Association and the Conference of representatives of Local Medical and Panel Committees declared that the measure of success which has attended the experiments of providing medical benefit under the National Health Insurance Act system has been sufficient to justify the profession in uniting to ensure the continuance and improvement of an Insurance system.

(a) Large numbers, indeed whole classes of persons are now receiving a real medical attention which they formerly did not receive at all.
(b) The number of practitioners in proportion to the population in densely populated areas has increased.
(c) The amount and character of the medical attention given is superior to that formerly given in the best of the old clubs, and immensely superior to that given in the great majority of the clubs which were far from the best.
(d) Illness is now coming under skilled observation and treatment at an earlier stage than was formerly the case.
(e) Speaking generally, the work of practitioners has been given a bias towards prevention which was formerly not so marked.
(f) Clinical records have been or are being provided which may be made of great service in relation to medical research and public health.
(g) Co-operation among practitioners is being encouraged to an increasing degree.
(h) There is now a more marked recognition than formerly of the collective responsibility of the profession to the community in respect of all health matters.

The association add that 'all these are immense gains, and though it is possible that some of them may not be wholly due to the establishment of the National Health Insurance Scheme, they have certainly been hastened and intensified by that system.'

The contract of the insurance practitioner with the insurance committee was subject to a number of regulations which were aimed at securing a certain standard of medical service and the necessary uniformity of procedure. It is fair to state that the doctor found no difficulty in following the requirements of these regulations as they affected his terms of service. For instance he must 'render all proper and necessary medical services other than those involving the application of special skill and experience of a degree or kind which general practitioners as a class cannot reasonably be expected to possess'. Also he must 'provide proper and sufficient surgery and waiting room accommodation', must 'order on a form provided for the purpose such drugs and prescribed appliances as are requisite for the treatment of the patient', must 'keep and furnish records of the diseases of his patients and of his treatment' and 'must give certificates to insured persons he is attending if they are incapable of work'. Of these requirements the keeping of records, prescribing and certification of incapacity for work are of particular interest.

A practitioner must keep records 'in such form as the Minister may from time to time determine'. An inter-departmental committee under the chairmanship of Sir Humphrey Rolleston issued a report in 1920 on which the form of record card was based. The objects of the record were clinical, administrative and statistical. The clinical notes were on an average kept satisfactorily by the majority of practitioners because of their value in diagnosis and treatment. A record has also to be made of each attendance and of occasions when a first and final certificate of incapacity were given. To many doctors in busy practices this obligation presented great difficulties and the records were not in fact used for statistical purposes.

Except in rural districts where there was no convenient pharmacy, patients obtained medicine from a chemist on a prescription issued by the doctor. In this way the scheme did much to transfer the dispensing of medicine from the doctor's surgery to the qualified chemist. There was no limit to the cost of medicines which a doctor might prescribe, though he was expected to exercise reasonable economy having regard to the therapeutic needs of his patient. In the early days of the Act many panel committees compiled formularies for the convenience of doctors on writing prescriptions and this led to some confusion when a prescription was dispensed in an area in which another formulary was in operation. To overcome this difficulty the Insurance Acts Committee of the B.M.A. compiled a national formulary which also included much useful information for the guidance of the practitioner and, in particular, a list of proprietary preparations with their therapeutic equivalents. The use of this formulary was widely adopted although it was made clear that 'it affected in no way the right of the medical practitioner to order such proper and sufficient medicines for his patients as the terms of his agreement with the Insurance Committee require'. There is no wish to stereotype prescribing and any practitioner can write extemporaneous prescriptions whenever he desires independently of the National Formulary. During the lifetime of the scheme many notable advances were made in therapeutics, and particularly in the treatment of the deficiency diseases. The insured person requiring prolonged treatment with such substances as insulin and liver extract could obtain them free of charge and without a constant drain on the family budget.

The question of issuing certificates of incapacity for work was one of great importance. It had two aspects. The first was economic, in ensuring that sickness benefit was paid to those patients only who by reason of their physical or mental condition were entitled to it. One important repercussion on the medical services was that the more money distributed in sickness benefit, the less was available for additional benefits. The second aspect was the harmful effect of inactivity in a patient who had recovered sufficiently to enable him to resume his normal activities. The doctor in assessing a case was frequently faced with problems requiring great judgment and on which there was, inevitably. some doubt. The approved societies, who were primarily concerned with the economic aspect, realised the necessity for some independent referee service available both to themselves and the practitioner. As a result the Regional Medical Service was established in 1920 as a department of the Ministry of Health. The decision of the regional medical officer was advisory only, and, when an opinion was given that a patient was fit for work, subsequent action might depend on whether the doctor issued any further certificates of incapacity. If he did so a second reference was usually made and, wherever it could possibly be arranged, the regional medical officer met the doctor in consultation. The need for a referee service of this kind was proved by the figures given in previous Reports. Suffice it to say that of patients referred by approved societies roughly 50 per cent. ceased drawing sickness benefit following the date of the reference. There were, of course, a number of patients who would have 'signed off' in any event, but of patients who were actually examined by the regional medical staff roughly 25 per cent. were considered fit to resume work.

The regional medical staff also performed other duties on behalf of the Ministry in connection with routine inspection of record cards and enquiries into allegations of excessive prescribing. By regular visits they were able to maintain a contact with practitioners which was of great value in helping the service to run with the minimum of friction.

If the National Health Insurance Act is regarded as an experiment in medical organisation on a nation wide basis its success can be judged by the fact that the incorporation of the general practitioner arrangements into the National Health Service was made with general approval of the profession. The new health service meant to a large number of practitioners simply an extension of their list to include not only the insured patients, but practically the whole of their practice.

In conclusion it should also not be forgotten that the life time of the National Health Insurance Scheme covered the period of the first and second world wars. During these critical years the welfare of industrial and agricultural workers was a matter of vital importance. The medical service supplied by the scheme to these workers represents a contribution the value of which cannot be over estimated.

7.5   H. H. Goldthorpe *et al.*, *Problems Arising from the Disposal of Effluents Containing Synthetic Detergents*, 1949

Sewage (and trade waste) may be defined as the water-borne wastes of a community. The sewage works can then be considered as the

Source: H. H. Goldthorpe, W. H. Hillier, C. Lumb and A. S. C. Lawrence, 'Problems arising from the disposal of effluents containing synthetic detergents', *Chemistry and Industry*, October 1949, pp. 679–680.

unloading station whence the vehicle, water, is discharged to the stream. The principal aim of sewage treatment is to free the polluted waters from settleable solids, to prevent the silting up of water courses, and to remove organic matter, such that when the final effluent is discharged to the stream, the re-oxygenation capacity of the stream is able to maintain a level of dissolved oxygen sufficient to support aerobic aquatic life.

The degree of purification required differs in various parts of the country depending on the relative volumes of the effluent and the receiving waters. In the West Riding, where the rivers are small, the Royal Commission Standards have been adopted. A final effluent would be considered satisfactory by the West Riding Rivers Board if (1) the suspended solids were less than 3 parts by weight per 100,000 and (2) the amounts of dissolved oxygen absorbed from good quality tap water at 65° C. in 5 days were less than 2 parts by weight per 100,000.

One must bear in mind that the polluting matter carried down to a sewage works varies widely, from hour to hour, from day to day, and from works to works, and very little control can be exercised over the quality or quantity of the incoming material. To achieve a satisfactory final effluent in the main, two general processes are carried out at a sewage works, namely, sedimentation followed by a biological process. Sedimentation, whether aided by precipitates such as lime, 'alumino ferric,' sulphuric acid, etc., or not, removes the settleable solids. The subsequent biological processes, for example, percolating beds or some form of activated sludge process, remove the fine suspensoids, colloidal and dissolved polluting matters.

A secondary function of a sewage works is to deal with the unloaded materials, which are in the form of watery viscous sludges. The very aqueous, water-retaining, gelatinous sludges from the biological processes are the most troublesome and every attempt is made to unload as much material in its unaltered form in the earlier sedimentation tanks, thereby also reducing the work of unloading to be performed by the biological processes.

The shortage of fats for the manufacture of soaps has caused synthetic detergents to become more and more general both in the home and in industry. The unloading devices of a sewage works have been built up around the properties of soap as the traditional detergent. Thousands of these synthetic surface active agents are possible, though, at present, only two, namely, Teepol and Lissapol N, have attained widespread commercial application, but others are on their way. Each new detergent has slightly different properties and those in control of present sewage undertakings are apprehensive as to their effect when discharged in indiscriminate mixtures in the near future.

At present the synthetic detergents are used in concentrations of 0.1% or more in the kitchen sink or scouring bowl. One has only to observe the sink or bowl after draining to notice the absence of scum or curd and to learn how efficient they are as detergents. Although it has been stated that the average amount of detergent arriving at the sewage works may not exceed 8 parts per 100,000, in actual practice quantities of the order of 100–200 parts per 100,000 are used domestically and industrially. In a previous symposium it was demonstrated that 20 parts per 100,000 when added to sewage were able to prevent entirely the settlement of the precipitable solids at all ranges of hydrogen-ion concentration.

In the kitchen sink or scouring bowl with some slight excess of detergent over the requirements for good dispersal of the unwanted material, very stable and well protected dispersoids must be formed. In the passage from sink or bowl, down the sewers to the settling tanks, change of composition is possible.

In experiments performed by the addition of detergents to samples of sewage the changes may not be the same as those made in the sink or bowl. It has also been claimed that the synthetic detergents discharged will help to keep the sewers clean and free from an ever-growing deposit of grease, but it must be remembered that these detergents have wetting and penetrative powers, leaving sewers and exposed iron work unprotected by the water-repellent alkaline earth soap curd, particularly at the wind and water line.

Sedimentation

When the sewage arrives at the works, it is detained in large tanks holding 8 hr. or more of the daily dry-weather flow, where everything settleable in that time sinks to the bottom and is collected as primary sludge. Equalising and mixing take place; Brownian movement and bacterial action play no small part in the clumping of the finer particles to settleable size. Precipitants such as lime, 'alumino ferric,' sulphuric acid, etc., are often added to aid settlement. Normally, with soap used as detergent the protective layer surrounding the fine particles is quickly destroyed and with hard waters or precipitants the curd formed helps in coagulating the dispersed matters, causing them both to settle rapidly. Where synthetic detergents are used, the solubilized or dispersed matter is unaffected by acids or alkaline earth metals and the matters hitherto unloaded in the sedimentation tanks are passed forward to the biological plants as an increased load which may be as high as 30% for a domestic sewage and more than 100% for certain industrial sewages.

Grease or oil which enters the sewers is readily emulsified and its early removal as scum is prevented. This emulsion also passes forward to the biological processes. In the woollen textile districts this loss of primary sludge and the increased loading of the biological plants are regarded with alarm. Large capital sums have been laid out for the recovery of grease from primary sludges, not so much for profit-making, but as an economic means for dealing with the large volumes of primary sludges produced. The presence of stable synthetic detergents transfers this work of unloading to the biological processes, where grease and fats are partly transformed and so mingled with biological material as to render de-watering and grease removal extremely difficult if not impracticable.

Biological processes

The biological processes employed for the purification of the settled liquids are the percolating bed or activated sludge. In both processes a complex mixture of minute plant and animal life is allowed to attack the impurities under aerobic conditions. Very little work has been done on the effect of synthetic detergents on these processes. Since synthetic detergents are not destroyed in the settling tanks, it is suspected that surface tension, influencing the transfer of oxygen, and interfacial tension between the bacterial cell and water, will be affected.

Holroyd, of Dagenham, has reported that using liquorice powder and egg albumin, the factor a/v in Adeney's formula for the solution of oxygen in water has been favourably affected. Some authors claim that enzymes play an important part in the coagulation of sewage impurities in biological processes. One has only to try the addition of saponin to rennet to observe the failure of junket making and to become apprehensive of the effect of synthetic detergents on bacterial enzymes. Bile salt, a surface active agent secreted into the intestine for the transfer of fats through the wet membrane of the intestinal wall, is bactericidal to many organisms except those of intestinal origin.

The results of adding Lissapol N and Teepol to the crude domestic and textile sewage of Huddersfield (possibly already containing these reagents), precipitating with acid, separation of any deposited sludge, neutralizing the supernatant water and treating with activated sludge, are as follows:

	Average results in parts per 100,000 over three weeks. Oxygen absorbed from $N/80$ $KMnO_4$ solution in 4 hours at $80°$ F. O.A.		20 hr. aeration—fill and draw. Dissolved oxygen absorbed by samples in 5 days at $65°$ F. B.O.A.	
	Influent	Effluent	Influent	Effluent
Precipitated sewage alone	14.5	2.05	25.67	0.43
÷20 pts. Lissapol N	15.6	2.68	26.76	0.66
÷20 pts. Teepol	16.2	2.29	34.26	0.78

The presence of the synthetic detergents increases the load to be borne by the plant. The effluent from the experiment with Lissapol N has a high 4 hours' permanganate (O.A.) value. The effluents containing the synthetic detergents have larger B.O.D. values than with sewage alone and Teepol gives the highest results.

Examination of the sludges from time to time revealed that the sludge treating sewage alone clotted readily and was full of infusorians. Lissapol N reduced the numbers of infusorians and only gave a fair sludge. Teepol produced a sludge in poor condition with few visible motile forms of life except under the high power of the microscope. The fear felt by those in control of sewage works is, that many present works are designed to produce a satisfactory effluent with very little margin of safety. As the curve of purification approaches asymptotically the line of complete purification, the presence of these synthetic detergents may require considerable extensions to the biological plants to deal not only with the increased load, but also with the more prolonged treatment necessary to arrive at the required B.O.D. of 2.0 parts per 100,000.

There is also the consideration of the secondary function of a sewage works, namely, that of dealing with the unloaded material. On the capability of a sewage works to deal with this unloaded material often depends the proper functioning of the works. In the West Riding towns this is a very important factor.

Only two synthetic detergents are at present in general use, but other large combines are formulating newer and possibly better compounds. If one considers the action of a detergent as the envelopment of the foreign matter in small locked compartments for conveyance to the sewage works, then those in control of sewage works must have some key at hand for unlocking and discharging those compartments. It would appear that with each new detergent a new key is required and keys of a very complicated pattern may be needed with indiscriminate mixtures. Such difficulties should be fully explored before an attempt at marketing is made.

8
Social and Human Engineering

8.1 F. W. Taylor, *Shop Management*, 1903

[. . .]

Modern engineering can almost be called an exact science; each year removes it further from guess work and from rule-of-thumb methods and establishes it more firmly upon the foundation of fixed principles.

The writer feels that management is also destined to become more of an art, and that many of the elements which are now believed to be outside the field of exact knowledge will soon be standardized, tabulated, accepted, and used, as are now many of the elements of engineering. Management will be studied as an art and will rest upon well recognized, clearly defined, and fixed principles instead of depending upon more or less hazy ideas received from a limited observation of the few organizations with which the individual may have come in contact. There will, of course, be various successful types, and the application of the underlying principles must be modified to suit each particular case. The writer has already indicated that he thinks the first object in management is to unite high wages with a low labor cost. He believes that this object can be most easily attained by the application of the following principles:

(a) A large daily task. Each man in the establishment, high or low, should daily have a clearly defined task laid out before him. This task should not in the least degree be vague nor indefinite, but should be circumscribed carefully and completely, and should not be easy to accomplish.

(b) Standard conditions. Each man's task should call for a full day's work, and at the same time the workman should be given such standardized conditions and appliances as will enable him to accomplish his task with certainty.

(c) High pay for success. He should be sure of large pay when he accomplishes his task.

(d) Loss in case of failure. When he fails he should be sure that sooner or later he will be the loser by it.

When an establishment has reached an advanced state of organization, in many cases a fifth element should be added, namely: the task should be made so difficult that it can only be accomplished by a first-class man.

There is nothing new nor startling about any of these principles and yet it will be difficult to find a shop in which they are not daily violated over and over again. They call, however, for a greater departure from the ordinary types of organization than would at first appear. In the case, for instance, of a machine shop doing miscellaneous work, in order to assign daily to each man a carefully measured task, a special planning department

Source: F. W. Taylor, 'Shop management', 1903, reprinted in *Scientific Management*, Westport, Conn.: Greenwood Press, 1972, pp. 63–66.

is required to lay out all of the work at least one day ahead. All orders must be given to the men in detail in writing; and in order to lay out the next day's work and plan the entire progress of work through the shop, daily returns must be made by the men to the planning department in writing, showing just what has been done. Before each casting or forging arrives in the shop the exact route which it is to take from machine to machine should be laid out. An instruction card for each operation must be written out stating in detail just how each operation on every piece of work is to be done and the time required to do it, the drawing number, any special tools, jigs, or appliances required, etc. Before the four principles above referred to can be successfully applied it is also necessary in most shops to make important physical changes. All of the small details in the shop, which are usually regarded as of little importance and are left to be regulated according to the individual taste of the workman, or, at best, of the foreman, must be thoroughly and carefully standardized; such details, for instance, as the care and tightening of the belts; the exact shape and quality of each cutting tool; the establishment of a complete tool room from which properly ground tools, as well as jigs, templets, drawings, etc., are issued under a good check system, etc.; and as a matter of importance (in fact, as the foundation of scientific management) an accurate study of unit times must be made by one or more men connected with the planning department, and each machine tool must be standardized and a table or slide rule constructed for it showing how to run it to the best advantage.

At first view the running of a planning department, together with the other innovations, would appear to involve a large amount of additional work and expense, and the most natural question would be is whether the increased efficiency of the shop more than offsets this outlay? It must be borne in mind, however, that, with the exception of the study of unit times, there is hardly a single item of work done in the planning department which is not already being done in the shop. Establishing a planning department merely concentrates the planning and much other brainwork in a few men especially fitted for their task and trained in their especial lines, instead of having it done, as heretofore, in most cases by high priced mechanics, well fitted to work at their trades, but poorly trained for work more or less clerical in its nature.

· There is a close analogy between the methods of modern engineering and this type of management. Engineering now centers in the drafting room as modern management does in the planning department. The new style engineering has all the appearance of complication and extravagance, with its multitude of drawings; the amount of study and work which is put into each detail; and its corps of draftsmen, all of whom would be sneered at by the old engineer as 'non-producers.' For the same reason, modern management, with its minute time study and a managing department in which each operation is carefully planned, with its many written orders and its apparent red tape, looks like a waste of money; while the ordinary management in which the planning is mainly done by the workmen themselves, with the help of one or two foremen, seems simple and economical in the extreme.

[. . .]

8.2   C Burt, *Analysis of Qualifications Needed for Particular Occupations*, 1923

Vocational guidance obviously implies a study not only of the character-istics of the individuals to be guided, but also of the characteristics of the employments they are to be guided into.

(a) The first step would seem to be an inventory of the various employ-ments available in the district in which the work of vocational guidance is to be undertaken [. . .] If more boys enter engineering trades than take up any other employment, if more girls become dressmakers or shorthand typists than anything else, these are the first occupations which call for 'job-analysis.'
'job-analysis.'

It may prove that the commonest jobs are sometimes those which have the fewest psychological limitations. To say that 'for the majority of men it is indifferent, from a psychological standpoint, into what employment they are directed' is, to my mind, too sweeping. But, certainly, since most jobs require only average abilities, and since most people have average abilities and no more, the task of the psychologist must often be a purely negative one—namely, demon-strating that a given individual is neither subnormal nor supernormal. And it becomes especially necessary—by a study of labour turnover, accidents, differences of output, and so forth—to find, first of all, what employments call for very special qualifications.

(b) The following-up of persons (particularly of persons already studied by psychological methods) after they have entered the occupations for which they have been recommended, will of itself throw valuable light upon the qualities needed in those particular employments. [. . .] A small intensive study could then be made of each group with a view to discovering which of the known qualities possessed by the various individuals had been favourable or unfavourable to success.

(c) It is desirable, however, to make, upon a more extensive scale, a scientific study of the qualities required for the commoner trades. The obvious procedure is to apply appropriate tests to large groups of actual workers already engaged; and to correlate the results of the tests with the independent judgments of the foreman or of the manager of the works. With this plan the chief obstacle is the difficulty of obtaining independent judgments which are unbiased and reliable. In some cases, where it is not possible for the firm to rank every individual in order of merit, it may prove easier for it to pick out, on the one hand, workers known to be highly efficient, and, on the other hand, those known to be highly inefficient; correspondence with the test-results will then be measured. [. . .]

(d) For a thorough and intensive study of the needs for particular trade processes, probably the best method will always be for a trained psychologist to seek temporary engagement as a worker

Source: C. Burt, 'The principles of vocational guidance', *The Journal of the National Institute of Industrial Psychology*, 1923, pp. 311–318.

himself, and so study by introspection the nature of the processes and capacities required.

(e) To do this for every possible trade and trade-process is a task that would demand an infinite allowance of time. For the urgent needs of the moment, it would seem wiser (i) to choose first of all those occupations for which the largest number of pupils enter, and the largest proportion of employees give dissatisfaction, and (ii) those occupations which are most closely akin to the most easily observable or measurable activities.

What qualities are to be assessed in making recommendations for vocational guidance?

(a) *Physical and medical.* The assessment of physical and medical qualifications or disabilities should present small difficulty. [. . .]

(b) *Educational attainments.* Educational attainments can now be measured with a fair degree of exactitude; and are of great importance for a certain limited group of occupations. [. . .] It has become highly necessary that attainments should be measured by means of a scientific scale of scholastic tests.

(c) Of all psychological capacities, general intelligence is the easiest to measure; it is also the most significant for the purposes of vocational guidance. Hitherto psychological investigation has succeeded in discovering no other general factor, certainly no other general factor with anything like so wide an applicability. [. . .] A certain degree of general intelligence, differing in different instances, is needful—and in theory assignable—for almost every kind of industrial work. More particularly, by its very nature, general intelligence seems closely related to the power to learn—to the ability to adapt oneself quickly and efficiently to new tasks and situations. Hence, for any trade which does not demand mere repetition of routine movements, or the mere reiterated application of knowledge and skill already acquired, general intelligence must be fundamental. In view, therefore, of the present difficulty of measuring other mental qualities, I am inclined to suggest that for the time being the chief (though not the sole) provisional guide in making vocational recommendations should be the degree of the candidate's intelligence.

Here two obvious principles will probably command almost an immediate—indeed, almost too ready an acceptance. (a) First, as a principle of guidance, other things being equal, a candidate should always be recommended to aim at the highest form of employment of which his intelligence is capable. (b) Secondly, as a principle of vocational selection, out of a given batch of candidates the most intelligent will always give the most satisfaction. And here it might be thought that the value of intelligence tests would end. In practice, however, certain important reservations must be added to each of these broad principles. In questions of vocational guidance, besides the candidate's general intelligence, his interests, special qualifications and financial means will frequently still further restrict his range of choice. Secondly in questions of vocational selection, an employer does not sufficiently realise the importance of a complementary principle, namely, that it is possible to engage an employee who is too intelligent for his job, as well as one who is too dull. [. . .] It is probable that for most occupations there is an optimal range of intelligence with an upper limit as well as a lower. One of the main tasks, therefore, of vocational guidance will be this: to draft each child into a job which corresponds as precisely as possible with his own level of intelligence, and requires neither a greater nor a less amount of mental capacity than he actually possesses.

A survey of London school children indicates that the whole population might be divided according to their intelligence into eight sections or grades.

1 The highest section includes the few children with mental ratios above 150. These are rarely found in elementary schools; and are almost entirely confined to the families from the higher social and professional classes. Among the general population they probably number less than 1 per mille. They constitute the type of pupils who win not only scholarships to secondary schools, but also scholarships to universities and honours degrees.

2 Secondly, there are children whose mental ratio is over 130 but under 150. This group comprises the top 1 or 2 per cent. of the elementary school population. The majority of them later on win scholarships to secondary schools, but to no higher college or university.

3 Children with mental ratios between 115 and 130 are in London usually drafted to schools now designated 'central,' but known formerly as 'higher elementary.' This group comprises rather less than the next 10 per cent.

4 The next section is a large one; and includes all who rise above the mean level of the whole, but who do not attain a mental ratio of more than 115. It forms 38 or 39 per cent. of the total.

5 The fifth section is equally large; and includes all who fall somewhat below the mean level of the whole, but again by no more than 15 per cent. of their age. This similarly contains nearly 40 per cent. of the total; and together these two sections constitute the dense medium of moderate ability, nearly four-fifths of the whole.

6 The sixth section includes those whose mental ratio lies between 85 and 70. These correspond roughly with what are now known as the 'dull and backward.' In an industrial area like London or Birmingham they form rather over 10 per cent. of the whole school population.

7 School children whose mental ratio falls below 70 are usually during childhood certified as feeble-minded; and committed to a special school. On reaching the age of 16, however, the majority of them are in practice virtually de-certified, and pass out into the world to earn their own living, if they can, like other children. They constitute rather more than the bottom 1 per cent. of the total school population.

8 The last group comprises those who are ineducable and will remain for ever in an institution—imbeciles and idiots with a mental ratio below 50. They constitute about 1 per mille of the child population.

Is it possible to arrange the occupations of adults, by strata, as it were, in any corresponding series? A provisional answer to this question may be sought in three different directions. First of all, there are the data collected by the American War Department which tested during the war a vast number of recruits, and classified the resulting performances according to occupation. Secondly, there are in this country the data obtained through the testing of candidates for Civil Service posts; these examinations, carried out during the past two or three years, have now covered nearly 40,000 candidates; and the range of ability displayed is almost as wide as that which would be found in a small group of schools, ordinary, central and secondary. Thirdly, isolated adults, many of them falling above or below even this wide range, have been tested in this country for various purposes—a small but increasing number of them by the National Institute, definitely for vocational guidance.

The results so collected tend to show that adults may also be classified into eight parallel groups, corresponding with the grade of their vocational work.

1 There are, first of all, adults, for the most part highly educated, and either following the higher professional careers (University lecturers, lawyers, medical men), or holding the higher administrative posts either in business or in the service of the State. These have a mental ratio[1] of over 150—averaging about 165.

2 Secondly, there are educated adults with lower professional qualifications (e.g. elementary teachers), and clerks holding higher and more responsible posts. These have a mental ratio usually over 130, averaging about 140. The intelligence of this group thus corresponds roughly with that of the secondary school child.

3 Thirdly, there are clerks of a lower grade doing for the most part work of an intelligent, but moderately routine, character. Workers engaged on highly skilled labour have also an intelligence at least equal to this group, These have a mental ratio averaging about 125; and usually over 115. The intelligence of this group thus corresponds roughly with that of the central school child.

4 Those engaged upon skilled labour usually have a mental ratio over 100 but seldom rising above 115. To the same group belong many of the ordinary commercial posts—successful shopkeepers and trades-people on a small scale, and shop-assistants employed by larger firms.

5 Fifthly, there is the large mass of semi-skilled labour, whose mental ratio falls below 100 but seldom sinks below 85.

6 The sixth category comprises unskilled labour. This category resembles in general level, though in actual numbers it somewhat exceeds, the dull and backward type of child. The mental ratio of the members comprising it falls usually below 85.

7 Seventhly, casual labour of the poorer types, and the lowest class of domestic service and of rural labour, show an intelligence little if at all above that of the high-grade feeble-minded child, and a mental ratio which is often well below 70.

8 Finally, there are the defective adults certified for institutional care. The mental ratio of most of these falls below 50.

The foregoing parallel is put forward rather as a rough but concrete illustration than as a scientific deduction from an exhaustive survey. As enumerated above, the vocational categories overlap enormously; nor is there sufficient data as yet available to define with any exactitude the limits of each class. It would seem, however, eminently desirable, even with our present imperfect knowledge, to work out, for any district in which a broad scheme of vocational guidance is attempted, the percentage of occupations available under each category.

For London a rough calculation (based mainly upon Charles Booth's survey, corrected by later census figures) gives the following proportions:

	Proportion of the total population per cent.
Vocational category	
I. Higher Professional	0.1
II. Lower professional	3
III. Clerks and highly skilled workers	12
IV. Skilled workers. Most commercial positions	27
V. Semi-skilled labour. Poorest commercial positions	36
VI. Unskilled labour, etc.	19
VII. Casual labour, etc.	3
VIII. Imbeciles and idiots	0.2

It will be observed that the proportions for the vocational categories show a fair correspondence with those for the different intellectual categories among the school children. But the distribution is less symmetrical. The percentages in the lower groups are too large, and those in the higher groups too small, for a perfect correspondence. [. . .]

Note

1   With adults of more than average intelligence a mental ratio is not directly calculable, since hardly any test-scales provide, or could provide, mental ages above 16. I believe that a far more scientific measure of intelligence is one expressed in terms either of the standard deviation or of percentiles (which, with a normal distribution, are convertible units). For the sake of simplicity I have expressed my marking throughout in terms of a mental ratio, basing my calculation on the assumptions that an individual's mental ratio and percentile position are constant throughout life.

8.3   Gene Richard, *On the Assembly Line*, 1937

As I walk down the steps with many others, I am disturbed by the thought that the day is only beginning. I suddenly realize in one sensation that there is no escape. It is all unavoidably real and painful. How much energy I must expend to-day has been predetermined by my employer. I try to disregard the thought that is causing this nervous tension which will be with me throughout the hours. Around me I sense a similar reaction. It expresses itself in silence. Men are laughing insincerely. They are ashamed of their emotion. They would rather feel that they were at peace and not a part of this herd who can hide nothing of their day from each other.

The men wander quietly into their places. The shop is beautiful. Machines, blue steel, huge piles of stock. Interesting patterns of windows

Source: Gene Richard, 'On the assembly line', *Atlantic Monthly*, April 1937, pp. 424–428.

are darkened by the early hour. This is the impression one gets before he becomes a part of the thing. The beauty is perceivable then. The unbiased observer cannot relate it to the subjective outlook he later acquires.

There is a shrill note. It is impersonal, commanding, and it expresses the entire power which orders the wheels set in motion. The conveyor begins to move immediately. Mysteriously the men are in their places and at work. A man near me grasps the two handles of the air wrench he holds all day long. This is the extent of his operation. He leans forward to each nut as the machine does its work. One nut–two nuts—one motor. It is not necessary for him to change his position. The conveyor brings the next motor to him. One position, one job all day.

Noise is deafening: a roar of machines and the groaning and moaning of hoists; the constant *pssffft-pssffft* of the air hoses. One must shout to be heard. After a time the noise becomes a part of what is natural and goes unnoticed. It merely dulls for the time the particular sense of hearing.

Truss works next to me. We are breaking a man in. There is a lot of experimenting to find out how to divide the jobs so as to achieve the maximum of group efficiency. The job is new to me, too. We are putting fuel connections on the carburetor. Between Truss and me we do three men's work. We cannot keep up. Luckily we know enough not to take it out on each other. We cuss and work in a fit of nervousness. The nut which is supposed to be previously tightened for me won't screw down because it is a bit undersized. I try to tighten it with my fingers, but I keep slipping behind. I am losing my temper. The foreman and relief man have been filling in occasionally for the man who should be there. We just can't do it. Truss snaps out, 'Hell with 'em! Let their damn motors go by if we can't get 'em!' We work and mumble curses. I finally discover how to put my wrench in the hole in such a way as to bite it into the soft brass and twist the lock nut down to where I can get a wrench on it. My ingenuity works out how to save my fingers, but to my disgust is merely adding to the possibility of Truss and me doing the job without help.

'Watch your quality to-day, men,' says Sammy, the squat line foreman. We are working so fast I don't see how anyone can think of quality. The old fellow next to me seems to be having trouble keeping up. He is supposed to run in a bolt on a clamp that I straighten and tighten with a hand wrench. When he gets behind I get behind, too. I take his ratchet wrench and do the added operation myself. I do this to two or three motors and give it back. Finally I just keep the wrench and do the added operation myself. I'll get sore each time I'm put behind anyway, so, to guarantee my own peace, I assume the extra work. He looks at me with mild appreciation and I go on feeling that I have big enough shoulders to make it easier for him. At least I'm younger and he's probably quite tired.

Up in the lavatory I usually lean out the window for a breath of fresh air. The out-of-doors smells fresh and free and reminds me how different it was when I could be outside and away from all this overwhelming noise and steel structure. But I can't take more than two or three breaths, for I must hurry back to the call of my stimulated conscience.

Men about me are constantly cursing and talking filth. Something about the monotonous routine breaks down all restraint. The men in most cases have little in common, but they must talk. The work will not absorb the mind of the normal man, so they must think. The feeling of isolation here leads one to the assurance that his confidences will never escape. Truss, without a trace of conscience, speaks of his more intimate relations with his wife. We work on and on with spurts of conversation. Suddenly a man breaks forth with a mighty howl. Others follow. We set up a howling all over the shop. It is a relief, this howling.

As the long-anticipated whistle blows for lunch the men burst into the aisles. There is a rule: 'No running.' Some of the men have developed a lunch-hour walk which is hard to distinguish from a run.

I am sitting on the greasy floor of the lunchroom leaning my back against the rail at the head of the stairs. The lunchroom is a great hall with many tables for the men to eat on. At the top of the stairs is a series of cages wide enough for a man to pass through when he rings his clock card. Twenty or thirty clocks are ringing, *ding-dong, ding-dong*, steadily for half an hour before the men go down to work. The floor is black from the dirty shoes. Some men's shoes are so soaked with oil that the surfaces shine and ooze at each step. The general manner of dress is not neat. The average worker probably wears a pair of work pants or old pants and a blue, brown, or black work shirt. Some wear vests. An old vest will protect the shirt and make a man feel dressed. In the cooler weather the whole costume can be covered when leaving the shop. In many cases it is done in such a manner as to create the illusion that the man is dressed much better than he really is. He usually has an old hat which, although it has become worn and dirty from handling, still retains form. An old topcoat then serves to disguise the rest.

In spite of the poorly regulated lives of these men, many gain weight. There are a great number of big massive hulks. This creates the impression of power. But I seldom see a man with a well-proportioned body. Some have a high left shoulder while the right droops. Some have large gnarled hands, the fingers of which fail to respond readily. Many hands lack a finger here and there. Most of the older men have a larger amount of beef in the region of the buttocks than they need. A protruding belly is almost the rule with the men who have been here long. The stomach muscles become relaxed and deformed from standing long hours in one position. I wonder if these men can be healthy. I suspect that they all have some nature of illness. The prevalence of halitosis might be accounted for some way.

Some of these men develop a surprisingly self-important air as though they were not a part of the group. They flaunt their independence. It has had me fooled since I've been here. Their attitude is effective, yet I sense there is something in it that is off color. The place has robbed these men of their true capacities and denied them a life of growth; but it cannot force them to be humble. Their outward front expresses an ownership of all those things they haven't got. They do even the most menial jobs with an air of great responsibility.

The shrill whistle blows. Some men start. It works as well as a whip. There is a rustling of clothing, a dropping of feet, and a prayer-like flow of voices as we go down the stairs.

This afternoon I am transferred to the rod department. My job is to weigh one end of the rod and stripe it with paint according to the colors indicated on the scale. There are usually a few piles of rods beside each man. The men figure it looks better to work this way. I take a rod off the pile and throw it on the scale, which is so made that the rod will sit on two pegs. The color is posted on the indicator instead of the weight, so all the operator needs to know is one color from another. I then pick up another rod, and as I take the first one off I put the next one on. While the scale is coming to a rest I paint the small end of the rod in my hand with a stripe corresponding to its weight. No time is lost. One soon gets so he can take a rod off the scale before it comes to a rest and predict where it will stop. As a matter of fact, to paint 5000 rods a day this is almost necessary.

As I am painting the small end of the rod I realize that I am not conscious of what I am doing. My accuracy surprises me. I seldom make a mistake, yet I never have my mind on my work. Perhaps this is why I am able to obtain accuracy, because my subconscious is more capable of this monotony than my personality.

It is soon after lunch. Someone has heard someone who heard someone else say the line was going home at two-thirty. Gradually it becomes a subject of discussion. Karl says to the bearer of the news, 'You wouldn't kid me, would ya? Cause that's a dirty trick.' 'Well, I heard a guy ask a foreman,' he said. We all know a foreman doesn't usually know any more than anyone else, yet we wishfully take stock in the rumor. The spirit of some men rises. Two o'clock finally arrives and there is no word yet. Karl curses the fellow who started the rumor. We still have hope, however, because we hate to abandon any chance of such a pleasant anticipation. After two o'clock we lose spirit.

Sometimes my thoughts will not hold me down. I think about all the mean things I have done, and all the things about myself I disrespect. Or I grow angry at some person out of my past. My thoughts go on and torture me. They are thoughts which I am sure are not sane. I try to stop thinking them and find that I don't really want to. I want to think them through until they satisfy me, and hope they will not come back. They do—and the process begins all over again. I cannot think them through to any finality because my work is constantly bringing me back to consciousness. These days and hours are bad. Sometimes I can lick my anxieties and think more objective thoughts. When I have contact with outside interests I can live them through the long hours of the day. Some days I have two or three good topics for thought. Then I am at peace and will postpone each pleasant thought smugly and with anticipation. As a beginner, I would try to think how fast each period of the day would go. This is a hard thing to get any satisfaction from. The day is just so long, and one gets to be as good a time reckoner as a clock.

I find now that I can put my mind to use. I have gained one thing from this hell. I have learned discipline. I can concentrate for an hour on one subject. But my efforts are fast losing direction. I have lost contact with anything to think about.

To-day I am thinking, as usual, depressed thoughts. I have heard these thoughts, some of them, expressed before, but now I am feeling them from dire reality. I have worked long hours this week. Each day I go to work in the dark and leave in the dark. I have not seen daylight since Sunday, and it is Saturday afternoon. I feel strangely unimportant and insignificant. The experiences of the day have exposed my mode of existence in such a way that I see my relative position here too plainly and deeply for my own comfort. I realize how unimportant is personal worth here. When I come in the gate in the morning I throw off my personality and assume a personality which expresses the institution of which I am a part. The only personality expressed here is the personality of the employer, through those authorized to represent him. There is no market for one's personal quality. Any expression of my own individual self beyond the scope of my work is in bad taste.

When a man insinuates here by any action that he is an individual, he is made to feel that he is not only out of place but doing something dishonorable. One feels that even the time he spends in the lavatory is not a privilege but an imposition. He must hurry back because there are no men to spare. After hours on one operation I realize that the only personal thing required of me is just enough consciousness to operate my body as a machine. Any consciousness beyond that is a contribution to my discomfort and maladjustment. I am a unit of labor, and labor is cheap. There is no market or appreciation of my worth except my self-respect. I struggle to keep it. My mood is perhaps a result of a discussion over the bench with Glen. He says, 'No matter whatcha do, they gotcha licked.' It makes me depressed to see him take himself so cheaply. He is convinced of his lack of value here. I feel a sudden wave of fear that I might some day feel exactly as he does.

Some of the men are taking to horseplay. Horseplay among bench workers has less limitation than among line workers. The bosses are not intolerant of horseplay. It is a noticeable fact that they will tolerate it where they will deal severely with serious loafing. As we are working we are unexpectedly interrupted by the foreman. He steps up between Karl and me. While we stop work and look around, he starts slowly to pave the way for what is to be a bawling out. His Swedish accent drawls out:

'Now listen, fellas. I don't know whether anybody ever tolja this before or whether ya know how it looks from the outside [glides his fingers over the bench in pattern of self-justification], but I'm gonna tell ya now. Now I ain't kickin' on how much work yer gittin' out er how well yer doin' it. Yer gittin' out enough perductchin and yer work's fine; but whatcher doin' is shovin' a whole buncha rods down the bench in a hurry and then gangin' up an' talkin'. Now if any one a those big shots come down 'ere an' see one guy leanin' on the bench like this, another guy over here standin' around, some guys bunched up here, an' everything all goin' ta hell, they wonder what kinda buncha guys they got down here and a hellova man runnin' it. Now I been takin' a lot up there lately an' I ain't been sayin' nothin'. Now I don't want to be a —— ——, but if I hafta I will. Those guys been comin' down here lately an' I been hearin' about it. They're kickin' an' they got a kick comin'. So —— damn it; you fellas work with me an' —— damn it I'll work with you, 'cause—well—ya see how it is, doncha? I ain't kickin' about yer work, but what I wancha ta do is—work a little slower if ya hafta and a little steadier.'

We start back to work in silence. It leaves a bad taste and we feel as though we really had been falling down on the job. Later we see him making the rounds, so we feel at least it wasn't meant just for us. We slip into some pretty childish ruts sometimes. We are so completely dulled by our work that trivial and boyish pranks amuse us. We cuss and talk filth.

When four-thirty finally arrives we get word that we are working until six. We have all settled into sullen moods. No one has a thing to say. We are grieved at this regular policy of detaining us without consulting us. Karl is working seriously for some time and finally drops back on one foot and bellows: '—— damn it! I'm gettin' sick of this stuff. I guess we never will get out of here before daylight.' He grabs the nearest rod and slams it down on the bench. I am mad too, so I egg him on. We take it out on the most faithful man in the department. Later we take to hollering to build up a morale which will help us to lick the last hour. Finally we are walking out, punching our cards. Laughter is now sincere but weary. It is still dark on the outside. I am so dulled that I have gotten here without realizing it. I stop—ponder. I can't think where I parked my car: the morning was so long ago.

9

Technology Transferred

9.1   V. I. Lenin, *The Immediate Tasks of the Soviet Government*, 1918

In every socialist revolution, however—and consequently in the socialist revolution in Russia which we began on October 25, 1917—the principal task of the proletariat, and of the poor peasants which it leads, is the positive or constructive work of setting up an extremely intricate and delicate system of new organisational relationships extending to the planned production and distribution of the goods required for the existence of tens of millions of people. Such a revolution can be successfully carried out only if the majority of the population, and primarily the majority of the working people, engage in independent creative work as makers of history. Only if the proletariat and the poor peasants display sufficient class-consciousness, devotion to principle, self-sacrifice and perseverance, will the victory of the socialist revolution be assured. By creating a new, Soviet type of state, which gives the working and oppressed people the chance to take an active part in the independent building up of a new society, we solved only a small part of this difficult problem. The principal difficulty lies in the economic sphere, namely, the introduction of the strictest and universal accounting and control of the production and distribution of goods, raising the productivity of labour and *socialising* production *in practice*. [. . .]

The new phase of the struggle against the bourgeoisie

The bourgeoisie in our country has been conquered, but it has not yet been uprooted, not yet destroyed, and not even utterly broken. That is why we are faced with a new and higher form of struggle against the bourgeoisie, the transition from the very simple task of further expropriating the capitalists to the much more complicated and difficult task of creating conditions in which it will be impossible for the bourgeoisie to exist, or for a new bourgeoisie to arise. Clearly, this task is immeasurably more significant than the previous one; and until it is fulfilled there will be no socialism.

If we measure our revolution by the scale of West-European revolutions we shall find that at the present moment we are approximately at the level reached in 1793 and 1871. We can be legitimately proud of having risen to this level, and of having certainly, in one respect, advanced somewhat further, namely; we have decreed and introduced throughout Russia the highest *type* of state—Soviet power. Under no

Source: V. I. Lenin, 'The immediate tasks of the Soviet government', 1918; in *Collected Works*, vol. 27, Moscow: Progress Publishers, 1965, pp. 241–269.

circumstances, however, can we rest content with what we have achieved, because we have only just started the transition to socialism, we have *not yet* done the decisive thing in *this* respect.

The decisive thing is the organisation of the strictest and country-wide accounting and control of production and distribution of goods. And yet, we have *not yet* introduced accounting and control in those enterprises and in those branches and fields of economy which we have taken away from the bourgeoisie; and without this there can be no thought of achieving the second and equally essential material condition for introducing socialism, namely, raising the productivity of labour on a national scale.

That is why the present task could not be defined by the simple formula: continue the offensive against capital. Although we have certainly not finished off capital and although it is certainly necessary to continue the offensive against this enemy of the working people, such a formula would be inexact, would not be concrete, would not take into account the *peculiarity* of the present situation in which, in order to go on advancing successfully *in the future*, we must 'suspend' our offensive *now*.

This can be explained by comparing our position in the war against capital with the position of a victorious army that has captured, say, a half or two-thirds of the enemy's territory and is compelled to halt in order to muster its forces, to replenish its supplies of munitions, repair and reinforce the lines of communication, build new storehouses, bring up new reserves, etc. To suspend the offensive of a victorious army under such conditions is necessary precisely in order to gain the rest of the enemy's territory, i.e., in order to achieve complete victory. Those who have failed to understand that the objective state of affairs at the present moment dictates to us precisely such a 'suspension' of the offensive against capital have failed to understand anything at all about the present political situation.

It goes without saying that we can speak about the 'suspension' of the offensive against capital only in quotation marks, i.e., only metaphorically. In ordinary war, a general order can be issued to stop the offensive, the advance can actually be stopped. In the war against capital, however, the advance cannot be stopped, and there can be no thought of our abandoning the further expropriation of capital. What we are discussing is the shifting of the *centre of gravity* of our economic and political work. Up to now measures for the direct expropriation of the expropriators were *in the forefront*. Now the organisation of accounting and control in those enterprises in which the capitalists have already been expropriated, and in all other enterprises, advances *to the forefront*.

If we decided to continue to expropriate capital at the same rate at which we have been doing it up to now, we should certainly suffer defeat, because our work of organising proletarian accounting and control has obviously—obviously to every thinking person—*fallen behind* the work of *directly* 'expropriating the expropriators'. If we now concentrate all our efforts on the organisation of accounting and control, we shall be able to solve this problem, we shall be able to make up for lost time, we shall *completely* win our 'campaign' against capital.

But is not the admission that we must make up for lost time tantamount to admission of some kind of error? Not in the least. Take another military example. If it is possible to defeat and push back the enemy merely with detachments of light cavalry, it should be done. But if this can be done successfully only up to a certain point, then it is quite conceivable that when this point has been reached, it will be necessary to bring up heavy artillery. By admitting that it is now necessary to make up for lost time in bringing up heavy artillery, we do not admit that the successful cavalry attack was a mistake. [. . .]

This is a peculiar epoch, or rather stage of development, and in order to defeat capital completely, we must be able to adapt the forms of our struggle to the peculiar conditions of this stage.

Without the guidance of experts in the various fields of knowledge, technology and experience, the transition to socialism will be impossible, because socialism calls for a conscious mass advance to greater productivity of labour compared with capitalism, and on the basis achieved by capitalism, Socialism must achieve this advance *in its own way*, by its own methods—or, to put it more concretely, by *Soviet* methods. And the specialists, because of the whole social environment which made them specialists, are, in the main, inevitably bourgeois. Had our proletariat, after capturing power, quickly solved the problem of accounting, control and organisation on a national scale (which was impossible owing to the war and Russia's backwardness), then we, after breaking the sabotage, would also have completely subordinated these bourgeois experts to ourselves by means of universal accounting and control. Owing to the considerable 'delay' in introducing accounting and control generally, we, although we have managed to conquer sabotage, have *not yet* created the conditions which would place the bourgeois specialists at our disposal. The mass of saboteurs are 'going to work', but the best organisers and the top experts can be utilised by the state either in the old way, in the bourgeois way (i.e., for high salaries), or in the new way, in the proletarian way (i.e., for creating the conditions of national accounting and control from below, which would inevitably and of itself subordinate the experts and enlist them for our work).

Now we have to resort to the old bourgeois method and to agree to pay a very high price for the 'services' of the top bourgeois experts. All those who are familiar with the subject appreciate this, but not all ponder over the significance of this measure being adopted by the proletarian state. Clearly, this measure is a compromise, a departure from the principles of the Paris commune and of every proletarian power, which call for the reduction of all salaries to the level of the wages of the average worker, which urge that careerism be fought not merely in words, but in deeds.

Moreover, it is clear that this measure not only implies the cessation—in a certain field and to a certain degree—of the offensive against capital (for capital is not a sum of money, but a definite social relation); it is also *a step backward* on the part of our socialist Soviet state power, which from the very outset proclaimed and pursued the policy of reducing high salaries to the level of the wages of the average worker.

Of course, the lackeys of the bourgeoisie, [. . .] will giggle over our confession that we are taking a step backward. But we need not mind their giggling. We must study the specific features of the extremely difficult and new path to socialism without concealing our mistakes and weaknesses, and try to be prompt in doing what has been left undone. To conceal from the people the fact that the enlistment of bourgeois experts by means of extremely high salaries is a retreat from the principles of the Paris Commune would be sinking to the level of bourgeois politicians and deceiving the people. Frankly explaining how and why we took this step backward, and then publicly discussing what means are available for making up for lost time, means educating the people and learning from experience, learning together with the people how to build socialism. There is hardly a single victorious military campaign in history in which the victor did not commit certain mistakes, suffer partial reverses, temporarily yield something and in some places retreat. The 'campaign' which we have undertaken against capitalism is a million times more difficult than the most difficult military campaign, and it would be silly and disgraceful to give way to despondency because of particular and partial retreat.

We shall now discuss the question from the practical point of view. Let us assume that the Russian Soviet Republic requires one thousand first-class scientists and experts in various fields of knowledge, technology and practical experience to direct the labour of the people towards securing the speediest possible economic revival. Let us assume also that we shall have to pay these 'stars of the first magnitude'—of course the majority of those who shout loudest about the corruption of the workers are themselves utterly corrupted by bourgeois morals—25,000 rubles per annum each. Let us assume that this sum (25,000,000 rubles) will have to be doubled (assuming that we have to pay bonuses for particularly successful and rapid fulfilment of the most important organisational and technical tasks), or even quadrupled (assuming that we have to enlist several hundred foreign specialists, who are more demanding). The question is, would the annual expenditure of fifty or a hundred million rubles by the Soviet Republic for the purpose of reorganising the labour of the people on modern scientific and technological lines be excessive or too heavy? Of course not. The overwhelming majority of the class-conscious workers and peasants will approve of this expenditure because they know from practical experience that our backwardness causes us to lose thousands of millions, and that we have *not yet* reached that degree of organisation, accounting and control which would induce all the 'stars' of the bourgeois intelligentsia to participate voluntarily in *our* work.

It goes without saying that this question has another side to it. The corrupting influence of high salaries—both upon the Soviet authorities (especially since the revolution occurred so rapidly that it was impossible to prevent a certain number of adventurers and rogues from getting into positions of authority, and they, together with a number of inept or dishonest commissars, would not be averse to becoming 'star' embezzlers of state funds) and upon the mass of the workers—is indisputable. Every thinking and honest worker and poor peasant, however, will agree with us, will admit, that we cannot immediately rid ourselves of the evil legacy of capitalism, and that we can liberate the Soviet Republic from the duty of paying an annual 'tribute' of fifty million or one hundred million rubles (a tribute for our own backwardness in organising *country-wide* accounting and control *from below*) only by organising ourselves, by tightening up discipline in our own ranks, by purging our ranks of all those who are 'preserving the legacy of capitalism', who 'follow the traditions of capitalism', i.e., of idlers, parasites and embezzlers of state funds (now all the land, all the factories and all the railways are the 'state funds' of the Soviet Republic). If the class-conscious advanced workers and poor peasants manage with the aid of the Soviet institutions to organise, become disciplined, pull themselves together, create powerful labour discipline in the course of one year, then in a year's time we shall throw off this 'tribute', which can be reduced even before that . . . in exact proportion to the successes we achieve in our workers' and peasants' labour discipline and organisation. The sooner we ourselves, workers and peasants, learn the best labour discipline and the most modern technique of labour, using the bourgeois experts to teach us, the sooner we shall liberate ourselves from any 'tribute' to these specialists. [. . .]

In order to proceed with the nationalisation of the banks and to go on steadfastly towards transforming the banks into nodal points of public accounting under socialism, we must first of all, and above all, achieve real success in increasing the number of branches of the People's Bank, in attracting deposits, in simplifying the paying in and withdrawal of deposits by the public, in abolishing queues, in catching and *shooting* bribe-takers and rogues, etc. At first we must really carry out the simplest things, properly organise what is available, and then prepare for the more intricate things.

Consolidate and improve the state monopolies (in grain, leather, etc.) which have already been introduced, and by doing so prepare for the state monopoly of foreign trade. Without this monopoly we shall not be able to 'free ourselves' from foreign capital by paying 'tribute'. And the possibility of building up socialism depends entirely upon whether we shall be able, by paying a certain tribute to foreign capital during a certain transitional period, to safeguard our internal economic independence.

We are also lagging very far behind in regard to the collection of taxes generally, and of the property and income tax in particular. The imposing of indemnities upon the bourgeoisie—a measure which in principle is absolutely permissible and deserves proletarian approval—shows that in this respect we are still nearer to the methods of warfare (to win Russia from the rich for the poor) than to the methods of administration. In order to become stronger, however, and in order to be able to stand firmer on our feet, we must adopt the latter methods, we must substitute for the indemnities imposed upon the bourgeoisie the constant and regular collection of a property and income tax, which will bring a *greater* return to the proletarian state, and which calls for better organisation on our part and better accounting and control.

The fact that we are late in introducing compulsory labour service also shows that the work that is coming to the fore at the present time is precisely the preparatory organisational work that, on the one hand, will finally consolidate our gains and that, on the other, is necessary in order to prepare for the operation of 'surrounding' capital and compelling it to 'surrender'. We ought to begin introducing compulsory labour service immediately, but we must do so very gradually and circumspectly, testing every step by practical experience, and, of course, taking the first step by introducing compulsory labour service *for the rich*. The introduction of work and consumers' budget books for every bourgeois, including

every rural bourgeois, would be an important step towards completely 'surrounding' the enemy and towards the creation of a truly popular accounting and control of the production and distribution of goods.
[. . .]

Raising the productivity of labour

In every socialist revolution, after the proletariat has solved the problem of capturing power, and to the extent that the task of expropriating the expropriators and suppressing their resistance has been carried out in the main, there necessarily comes to the forefront the fundamental task of creating a social system superior to capitalism, namely, raising the productivity of labour, and in this connection (and for this purpose) securing better organisation of labour. Our Soviet state is precisely in the position where, thanks to the victories over the exploiters—from Kerensky to Kornilov—it is able to approach this task directly, to tackle it in earnest. And here it becomes immediately clear that while it is possible to take over the central government in a few days, while it is possible to suppress the military resistance (and sabotage) of the exploiters even in different parts of a great country in a few weeks, the capital solution of the problem of raising the productivity of labour requires, at all events (particularly after a most terrible and devastating war), several years. The protracted nature of the work is certainly dictated by objective circumstances.

The raising of the productivity of labour first of all requires that the material basis of large-scale industry shall be assured, namely, the development of the production of fuel, iron, the engineering and chemical industries. The Russian Soviet Republic enjoys the favourable position of having at its command, even after the Brest peace, enormous reserves of ore (in the Urals), fuel in Western Siberia (coal), in the Caucasus and the South-East (oil), in Central Russia (peat), enormous timber reserves, water power, raw materials for the chemical industry (Karabugaz), etc. The development of these natural resources by methods of modern technology will provide the basis for the unprecedented progress of the productive forces.

Another condition for raising the productivity of labour is, firstly, the raising of the educational and cultural level of the mass of the population. This is now taking place extremely rapidly, a fact which those who are blinded by bourgeois routine are unable to see; they are unable to understand what an urge towards enlightenment and initiative is now developing among the 'lower ranks' of the people thanks to the Soviet form of organisation. Secondly, a condition for economic revival is the raising of the working people's discipline, their skill, the effectiveness, the intensity of labour and its better organisation.

In this respect the situation is particularly bad and even hopeless if we are to believe those who have allowed themselves to be intimidated by the bourgeoisie or by those who are serving the bourgeoisie for their own ends. These people do not understand that there has not been, nor could there be, a revolution in which the supporters of the old system did not raise a howl about chaos, anarchy, etc. Naturally, among the people who have only just thrown off an unprecedentedly savage yoke there is deep and widespread seething and ferment; the working out of new principles of labour discipline by the people is a very protracted process, and this process could not even start until complete victory had been achieved over the landowners and the bourgeoisie.

We, however, without in the least yielding to the despair (it is often false despair) which is spread by the bourgeoisie and the bourgeois intellectuals (who have despaired of retaining their old privileges), must under no circumstances conceal an obvious evil. On the contrary, we shall expose it and intensify the Soviet methods of combating it, because the victory of socialism is inconceivable without the victory of proletarian conscious discipline over spontaneous petty-bourgeois anarchy, this real guarantee of a possible restoration of Kerenskyism and Kornilovism.

The more class-conscious vanguard of the Russian proletariat has already set itself the task of raising labour discipline. For example, both the Central Committee of the Metalworkers' Union and the Central Council of Trade Unions have begun to draft the necessary measures and decrees. This work must be supported and pushed ahead with all speed. We must raise the question of piece-work and apply and test it in practice; we must raise the question of applying much of what is scientific and progressive in the Taylor system; we must make wages correspond to the total amount of goods turned out, or to the amount of work done by the railways, the water transport system, etc., etc.

The Russian is a bad worker compared with people in advanced countries. It could not be otherwise under the tsarist regime and in view of the persistence of the hangover from serfdom. The task that the Soviet government must set the people in all its scope is—learn to work. The Taylor system, the last word of capitalism in this respect, like all capitalist progress, is a combination of the refined brutality of bourgeois exploitation and a number of the greatest scientific achievements in the field of analysing mechanical motions during work, the elimination of superfluous and awkward motions, the elaboration of correct methods of work, the introduction of the best system of accounting and control, etc. The Soviet Republic must at all costs adopt all that is valuable in the achievements of science and technology in this field. The possibility of building socialism depends exactly upon our success in combining the Soviet power and the Soviet organisation of administration with the up-to-date achievements of capitalism. We must organise in Russia the study and teaching of the Taylor system and systematically try it out and adapt it to our own ends. At the same time, in working to raise the productivity of labour, we must take into account the specific features of the transition period from capitalism to socialism, which, on the one hand, require that the foundations be laid of the socialist organisation of competition, and, on the other hand, require the use of compulsion, so that the slogan of the dictatorship of the proletariat shall not be desecrated by the practice of a lily-livered proletarian government.

The organisation of competition

[. . .]

In the rail transport service, which perhaps most strikingly embodies the economic ties of an organism created by large-scale capitalism, the struggle between the element of petty-bourgeois laxity and proletarian organisation is particularly evident. The 'administrative' elements provide a host of saboteurs and bribe-takers: the best part of the proletarian elements fight for discipline; but among both elements there are, of course, many waverers and 'weak' characters who are unable to withstand the 'temptation' of profiteering, bribery, personal gain obtained by spoiling the whole apparatus, upon the proper working of which the victory over famine and unemployment depends.

The struggle that has been developing around the recent decree on the management of the railways, the decree which grants individual executives dictatorial powers (or 'unlimited' powers), is characteristic. The conscious (and to a large extent, probably, unconscious) representatives of petty-bourgeois laxity would like to see in this granting of 'unlimited' (i.e., dictatorial) powers to individuals a departure from the collegiate principle, from democracy and from the principles of Soviet government. Here and there, among Left Socialist-Revolutionaries, a positively hooligan agitation, i.e., agitation appealing to the base instincts and to the small proprietor's urge to 'grab all he can', has been developed against the dictatorship decree. The question has become one of really enormous significance. Firstly, the question of principle, namely, is the appointment of individuals, dictators with unlimited powers, in general compatible with the fundamental principles of Soviet government? Secondly, what relation has this case—this precedent, if you will—to the special tasks of government in the present concrete situation? We must deal very thoroughly with both these questions.

That in the history of revolutionary movements the dictatorship of individuals was very often the expression, the vehicle, the channel of the dictatorship of the revolutionary classes has been shown by the irrefutable experience of history. Undoubtedly, the dictatorship of individuals was compatible with bourgeois democracy. On this point, however, the bourgeois denigrators of the Soviet system, as well as their petty-bourgeois henchmen, always display sleight of hand: on the one hand, they declare the Soviet system to be something absurd, anarchistic and savage, and carefully pass over in silence all our historical examples and theoretical arguments which prove that the Soviets are a higher form of democracy, and what is more, the beginning of a *socialist* form of democracy; on the other hand, they demand of us a higher democracy than bourgeois democracy and say: personal dictatorship is absolutely incompatible with your, Bolshevik (i.e., not bourgeois, *but socialist*), Soviet democracy.

These are exceedingly poor arguments. If we are not anarchists, we must admit that the state, *that is, coercion*, is necessary for the transition from capitalism to socialism. The form of coercion is determined by the degree of development of the given revolutionary class, and also by special circumstances, such as, for example, the legacy of a long and reactionary war and the forms of resistance put up by the bourgeoisie and the petty bourgeoisie. There is, therefore, absolutely *no* contradiction in principle between Soviet (*that is,* socialist) democracy and the exercise of dictatorial powers by individuals. The difference between proletarian dictatorship and bourgeois dictatorship is that the former strikes at the exploiting minority in the interests of the exploited majority, and that it is exercised—*also through individuals*—not only by the working and exploited people, but also by organisations which are built in such a way as to rouse these people to history-making activity. (The Soviet organisations are organisations of this kind.)

In regard to the second question, concerning the significance of individual dictatorial powers from the point of view of the specific tasks of the present moment, it must be said that large-scale machine industry—which is precisely the material source, the productive source, the foundation of socialism—calls for absolute and strict *unity of will*, which directs the joint labours of hundreds, thousands and tens of thousands of people. The technical, economic and historical necessity of this is obvious, and all those who have thought about socialism have always regarded it as one of the conditions of socialism. But how can strict unity of will be ensured? By thousands subordinating their will to the will of one.

Given ideal class-consciousness and discipline on the part of those participating in the common work, this subordination would be something like the mild leadership of a conductor of an orchestra. It may assume the sharp forms of a dictatorship if ideal discipline and class-consciousness are lacking. But be that as it may, *unquestioning subordination* to a single will is absolutely necessary for the success of processes organised on the pattern of large-scale machine industry. On the railways it is twice and three times as necessary. In this transition from one political task to another, which *on the surface* is totally dissimilar to the first, lies the whole originality of the present situation. The revolution has only just smashed the oldest, strongest and heaviest of fetters, to which the people submitted under duress. That was yesterday. Today, however, the same revolution demands—precisely in the interests of its development and consolidation, precisely in the interests of socialism—that the people *unquestioningly obey the single will* of the leaders of labour. Of course, such a transition cannot be made at one step. Clearly, it can be achieved only as a result of tremendous jolts, shocks, reversions to old ways, the enormous exertion of effort on the part of the proletarian vanguard, which is leading the people to the new ways. [. . .]

Notes on Authors

Baekeland, Leo Hendrik (1863–1944). Born in Ghent, Baekeland, trained as an organic chemist at the University there (PhD 1884). He emigrated to the US and worked as an industrial consultant, inventing a photographic paper which earned him his first fortune. The first of the 'bakelite' patents was taken out in 1906. A German firm was set up for its manufacture in 1910, and Baekeland set up his own firm in New Jersey in 1922.

Belloc, Hilaire (1870–1953). An Anglo-French historian, essayist, novelist, poet and children's writer, Belloc graduated in history from Oxford and settled in Sussex. He was a political individualist with a reputation as a controversialist, strongly opposed to free thought and socialism. He combined an intense Catholicism with a backward-looking enthusiasm for the medieval period.

Burt, Sir Cyril (1883–1971). Famous for his views on the inheritance of intelligence and for establishing the importance of educational psychology, Burt studied at Oxford and was psychologist to the London County Council from 1913 to 1932. From 1932 to 1950, he held the chair of psychology at the University of London. A somewhat difficult personality in life, after his death, some of his most well-known work on intelligence and educational standards was shown incontrovertibly to have been based on fabricated evidence.

Chase, Stuart (1888–1985). An American economist and writer. Chase studied at Harvard. In the 1920s, he was a member of the Technical Alliance, a group focused around the ideas of Thorstein Veblen who believed that power should be turned over to 'technicians' who would administer a planned economy. Later, Chase worked for a non-profit organization providing technical research services for labour unions and cooperatives and established a career as a writer on economic and social subjects.

Craven, T.A.M. (1893–1972). Radio engineer; Chief Engineer (1935–1937), Engineering Department, Federal Communications Commission, Washington DC; Commissioner (1937–1944).

Electrical Association for Women (1924–1986). This organization was founded by women engineers anxious to ease women's domestic burdens through the intelligent use of electricity. They aimed both to educate women about the role of technology in the home and to serve as a forum for bringing women's views on domestic technology to the attention of the industry. The EAW published *The Electrical Age [for Women]*.

Ewing, Sir Alfred (1855–1935). After studying engineering and physics at Edinburgh, Ewing pursued a largely academic career, including a spell modernizing naval education from 1903 to 1916 to include practical engineering training, which he believed would promote efficiency. During the Second World War, he played an important role in code-breaking. From 1916 to 1929, he was principal and vice-chancellor of Edinburgh University where he instituted the PhD degree, and brought about a large building programme for the sciences and engineering.

Ferranti, Sebastian Ziani de (1864–1930). An electrical engineer who trained in the nascent industry, Ferranti was an early proponent of large-scale central generation and high voltage AC transmission. An early scheme for a large London power station foundered on technical and legal problems and he became an electrical manufacturer. He maintained his advocacy of large-scale central generation.

Florey, Howard W. (1898–1968). Trained initially as a doctor in Adelaide and then as a physiologist at Oxford and Cambridge, Florey turned to experimental pathology, holding a succession of distinguished posts including the chair of pathology at Oxford from 1934. He is most famous as the creator of penicillin therapy during the early years of the Second World War. Failing to get commercial assistance in Britain, he turned his laboratory into a factory and then sought commercial sponsorship and patents in the U.S., which caused some controversy as did a priority dispute over the use of penicillin with Alexander Fleming. He was later president of the Royal Society of London and provost of Queen's College, Oxford.

Giedion, Siegfried (1893–1968). The son of a Swiss manufacturer, Giedon studied both engineering and art history. He was interested particularly in the history of architecture which he taught at Harvard and later in Zurich.

Goldthorpe, Harold H. BSc (Lond) DSc (Strasburg). Chemist and Bacteriologist, Sewage and Water Works, Huddersfield Corporation.

Hodge, John (1855–1937). From humble beginnings, Hodge eventually came to play a leading role in fostering unionization in the iron and steel industry in Britain. He was also active in political and social movements, serving as a Labour MP from 1906–1923, with wartime spells as Minister for Labour and Minister for Pensions. He played a major part in the formation of Conciliation Boards for the prevention of industrial disputes.

Lancaster, Maud (born 1870). Probably related to the civil and electrical engineer mentioned on her title page.

Lenin, V.I. (1870–1924). Leader of the Bolshevik Revolution in 1917, Lenin attempted to carry through a programme of economic revolution, including heavy investment in modern science-based technology, which he saw as the key to Russia's future.

Loasby, Geoffrey BSc (Birm), FRIC, FTI. Research and Production Manager, British Nylon Spinners Ltd.

Marvin, Francis Sydney (1863–1943). Marvin had a career as a teacher and later as an Inspector for the Board of Education. He also worked as an organizer of courses and lectures for teachers of modern languages and history.

Orr, Lord John Boyd (1880–1971). Trained as a physician and in physiology at Glasgow, Orr studied nutrition. His nutritional work had policy implications, which were often quashed. Because of it, he argued for such measures as free school milk in the 1920s. He drew attention to the poor state of health of the British people in the 1930s, and began to think on a global scale as well. He was the first director general of the UN Food and Agriculture Organization (FAO), but resigned after its rejection of his World Food Plan. In later years, he acted as advisor on food and agriculture matters to developing countries.

Osgood, Harold A. (c.1885–1940). Vice-President, Fulton Iron Works Company, St. Louis, Missouri, and authority on rail transport.

Potter, Howard V. (1888–1970). Chief Chemist, then Technical Director of the Bakelite Company.

Pratt, Edwin A. (1854–1922). Pratt wrote several books on transport history. He was especially interested in nationalization and the role of railways in wartime.

Reade, William Winwood (1838–1875). Reade was a prolific writer, enthusiast for science and traveller who made several exploratory expeditions into Africa.

Richard, Gene (born c.1913). An automobile worker, Richard had previously tried to be a professional musician.

Sand, René (1877–1953). Belgian founder of the International Hospitals Association.

Scott, Charles F. (1864–1944). He was an electrical engineer who studied at Ohio State and The Johns Hopkins University. Scott was employed by Westinghouse and became a leading exponent of the universal system of large-scale central generation of high-voltage AC electricity for long distance transmission and distribution to meet a variety of needs. He was active in developing the Niagara Falls project and later became professor of electrical engineering at Yale.

Simon, Sir John (1816–1904). Trained as a medical man and interested in pathology, Simon became the first Medical Officer for Health for London in 1848. He was an ardent sanitary reformer with administrative talent and his numerous reports were influential in government circles and in public health reform.

Taylor, Frank Sherwood (1897–1956). Sherwood Taylor began his career as an academic chemist, but then turned to the history of science eventually becoming Director of the Science Museum in London. He was worried in the 1930s about a widening gulf between the arts and the sciences, and used the history of science as a form of bridge. Much of his writing had a mystical theme. In later years he turned to Catholicism.

Taylor, Frederick Winslow (1856–1915). An American engineer, Taylor began his career in a steel mill, where he began the time and motion study of work processes which led to the development of his ideas on scientific management of production. His ideas were very influential in the early years of mass production.

White, Hugh (1876–1936). A leading American builder, President and Chairman of the Board, George A. Fuller Company.

Index